FOURTH EDITION

COMMUNICATION at WORK

MARY FINLAY
Seneca College

CHRISTINE FRANK
Georgian College

NELSON EDUCATION

NELSON EDUCATION

Communication at Work
Fourth Edition

by Mary Finlay and Christine Frank

Associate Vice-President, Editorial Director:
Evelyn Veitch

Publisher:
Joanna Cotton

Marketing Manager:
Wayne Morden

Developmental Editor:
Natalie Barrington

Production Editor:
Wendy Yano

Senior Production Coordinator:
Hedy Sellers

Copy Editor/Proofreader:
Wayne Herrington

Permissions Coordinator:
Cindy Howard

Indexer:
Elizabeth Bell

Creative Director:
Angela Cluer

Interior Design:
Tammy Gay

Cover Design:
Rocket Design

Cover Image:
Creatas/First Light

Compositor:
Nelson Gonzalez

COPYRIGHT © 2007 by Nelson Education Ltd.

Printed and bound in Canada
4 5 6 7   15 14 13 12

For more information contact Nelson Education Ltd., 1120 Birchmount Road, Toronto, Ontario, M1K 5G4. Or you can visit our Internet site at http://www.nelson.com

ALL RIGHTS RESERVED. No part of this work covered by the copyright herein may be reproduced, transcribed, or used in any form or by any means—graphic, electronic, or mechanical, including photocopying, recording, taping, Web distribution, or information storage and retrieval systems—without the written permission of the publisher.

For permission to use material from this text or product, submit all requests online at www.cengage.com/permissions. Further questions about permissions can be emailed to permissionrequest@cengage.com

Every effort has been made to trace ownership of all copyrighted material and to secure permission from copyright holders. In the event of any question arising as to the use of any material, we will be pleased to make the necessary corrections in future printings.

Library and Archives Canada Cataloguing in Publication

Finlay, Mary
Communication at work / Mary Finlay, Christine Frank.—4th ed.

Includes index.
ISBN 0-17-640703-0

1. Business communication—Textbooks. 2. Commercial correspondence—Textbooks. 3. Business report writing—Textbooks. I. Frank, Christine, 1951- II. Title.

HF5718.F55 2006   658.4'5
C2005-907158-3

PREFACE

TO THE INSTRUCTOR

Like its predecessors, this edition of *Communication at Work* is designed to help students apply the communication skills they have acquired in school, at work, and in their personal lives to speaking and writing situations that they will encounter in their careers after graduation. Examples drawn from likely work situations encourage students to discover and analyze what they already know about effective communication. The text organizes this analysis into concrete steps for approaching new situations and gives students a chance to apply these steps to new situations taken from a wide range of employment areas, thus providing immediate feedback on their understanding of the material.

The oral communication sections have been integrated into each chapter to emphasize the similarities between, for instance, relating good news orally and giving good news in writing. This approach recognizes the fact that most college graduates will use oral communication skills in meetings, interviews, telephone calls, and informal exchanges rather than in structured oral presentations. Similarly, communication technologies such as voice mail, e-mail, and facsimile machines are discussed throughout the text as tools that can help workers achieve their communication objectives more efficiently. This approach reinforces the important point that technology does not render good writing and speaking skills obsolete or even require workers to acquire new skills, but simply allows good writing and speaking to be applied with less mechanical effort. Removing many of the boring and repetitive elements of communicating on the job leaves time for planning and revising.

In addition, because oral communication within an organization is an important part of workplace communication, we have included a chapter that discusses such topics as oral presentations, working in groups, meetings, teleconferencing, and videoconferencing.

More than ever, this textbook has been designed specifically to help students understand the importance of good communication skills and how they apply to their job searches and careers. For this reason, we begin with a chart, "Employability Skills Profile," from The Conference Board of Canada. We have also included chapters on résumés, application letters, and interviews.

NEW TO THE FOURTH EDITION
- The major change in this edition is the focus on e-mail as a primary means of communication, reinforced by many new exercises and examples.
- The section on the informal report has also been revised to provide more step-by-step instruction.
- The section on presentations has been updated with more emphasis on presentation software and a new example that includes a graph.
- Chapter 12 has undergone considerable updating and expansion on research strategies and resources, especially via the Internet, and contains updated examples of APA and MLA documentation.
- A new sample formal report that reflects research using both primary and secondary resources and also contains graphs has been added. You will notice that the subject of the report matches the PowerPoint presentation in Chapter 8.
- Pedagogy has been added:
  - The "ESL Tips" provide extra assistance for students who are not native English speakers.
  - Each chapter contains a self-test for review of the main concepts.
  - Exercises have been updated, and new exercises have been added.
  - Icons have been included to illustrate "poor" and "improved" examples.
  - Cartoons have been added to the part openers.

INSTRUCTOR'S MANUAL

To help you prepare for classes, an Instructor's Manual (ISBN 0-17-625165-0) accompanies *Communication at Work*. You will find such useful topics as classroom activities, suggested answers to the exercises in the text, and transparency masters. We hope this manual will give you the necessary tools to enhance your students' understanding and learning.

TO THE STUDENT

Communication at Work is designed to help you apply your speaking and writing skills to the type of communication situations you are likely to encounter at work.

Whatever career you plan to pursue, the ability to communicate effectively will be important to your success. When employers are surveyed about the skills they look for in hiring and promoting personnel, communication skills always rank near the top of the list (see "Employability Skills Profile" from The Conference Board of Canada at the beginning of the book). While the technical expertise or subject content you acquire in other courses will quickly become obsolete, you will be able to apply the principles of good communication to whatever you do. Furthermore, your ability as a reader, writer, speaker, and listener will help you acquire new information and new technical skills that will keep you productive and motivated throughout your career.

To get the most out of your textbook, in conjunction with your classroom participation and take-home assignments, take the following steps:

1. When your instructor assigns a chapter, read the learning objectives on the first page to get an overview of the material.
2. The material that introduces major topics in each chapter is a chance for you to identify what you already know about effective writing and speaking at work. Read the cases carefully, and answer the questions in writing. Your instructor may ask you to do this in class. Compare your answers with the ones in the text, and note any significant differences. These steps will help you to integrate new insights with what you already know.
3. Sections of the text after the case studies break down general principles into practical step-by-step explanations. You will have frequent opportunities to practise these steps by doing the exercises in your book. Again, your teacher may ask you to work on these in class.
4. Reread the learning objectives. Each objective is an action that you should now be able to carry out.
5. Your instructor may assign one or more of the exercises in the textbook, or may give you an alternative assignment to test your success at achieving the objectives of the chapter.
6. Complete the Chapter Review/Self-Test at the end of each chapter to determine which topics you may need to spend more time on.

A NOTE ON THE DESIGN

We would like to point out a feature of this book that will help you with your studies. You will see that we have linked each chapter objective to a section title with the icon "■". The icon contains the relevant chapter objective number. We have specifically designed this so that you will know your learning objective or outcome for each section of the chapter. We hope you will find this a useful design element.

We have also added to this edition an "improved" ✓ and "poor" ✗ icon to help you quickly identify the examples.

Although the sample memos, letters, and reports have been designed to look "real," their sizes have been reduced to fit the pages of this book. What you see, therefore, are not actual or true sizes. Please ask your teacher or read through this book for the required specifications of such things as line spacing, margin spacing, and font size.

ACKNOWLEDGMENTS

Mary Finlay would like to thank her husband, William Riddell, for his support in preparing *Communication at Work*, Fourth Edition. Christine Frank likewise thanks her husband, David Melanson.

We owe many thanks to everyone who reviewed and provided ideas on improving this new edition: Liesje de Burger, University of Ontario Institute of Technology; Joan Flaherty, University of Guelph; Rhonda Hustler, Centennial College; Heather Larsen, Cambrian College; Richard McMaster, Ryerson University; Anne Price, Red Deer College; Don Roberts, Seneca College; Jean Timbury, Algonquin College; William Van Nest, Sir Sandford Fleming College.

Finally, the team at Thomson Nelson needs to be acknowledged for their tireless efforts: Mike Thompson, who decided this title had some life left in it, Natalie Barrington, Wendy Yano, and Wayne Herrington.

Mary Finlay
Christine Frank
August 2005

BRIEF CONTENTS

Employability Skills Profile   xiii

PART ONE   Introduction to Communication   1
Chapter One   The Communication Process   2
Chapter Two   Style in Business Communication   21

PART TWO   Communication Styles   53
Chapter Three   Presenting Written Messages   54
Chapter Four   Routine Messages   79
Chapter Five   Good News: The Direct Approach   107
Chapter Six   Bad News: The Indirect Approach   126
Chapter Seven   Persuasive Messages   146
Chapter Eight   Oral Communication within the Organization   175

PART THREE   The Job Search   199
Chapter Nine   Résumés   200
Chapter Ten   Locating and Applying for Jobs   226
Chapter Eleven   Interviews   247

PART FOUR   The Report   265
Chapter Twelve   Researching the Report   266
Chapter Thirteen   Planning the Informal Report   304
Chapter Fourteen   Presenting the Formal Report   330

APPENDIX   Handbook of Grammar and Usage   359

INDEX   401

CONTENTS

Employability Skills Profile   xiii

PART ONE   Introduction to Communication   1

Chapter One   The Communication Process   2

    Chapter Objectives   2
    Introduction: Three Parts of the Communication Process   3
    Identifying Your Purpose   4
    Identifying Your Audience   7
    Selecting the Appropriate Medium   8
    Nonverbal Messages   12
    Eliminating Communication Barriers   15
    Chapter Review/Self-Test   15
    Exercises   17

Chapter Two   Style in Business Communication   21

    Chapter Objectives   21
    Introduction   22
    Writing for Clarity   22
    Clarity in Oral Communication   30
    Tone   31
    Tone in Oral Communication   44
    Answers to Chapter Questions   46
    Chapter Review/Self-Test   47
    Exercises   48

PART TWO   Communication Styles   53

Chapter Three   Presenting Written Messages   54

    Chapter Objectives   54
    Introduction   55
    Letter Presentation   55

      Memorandum Presentation   68
      E-Mail Presentation   69
      Facsimile Message Presentation   71
      Chapter Review/Self-Test   72
      Exercises   74

Chapter Four   Routine Messages   79

      Chapter Objectives   79
      Introduction   80
      Making a Routine Request   80
      Placing an Order   85
      Routine Replies   89
      Oral Communication   95
      Answers to Chapter Questions   99
      Chapter Review/Self-Test   99
      Exercises   101

Chapter Five   Good News: The Direct Approach   107

      Chapter Objectives   107
      Introduction   108
      Writing Good News Messages   108
      Responding to Complaints or Problems   110
      Delivering Good News Orally   118
      Answers to Chapter Questions   120
      Chapter Review/Self-Test   120
      Exercises   122

Chapter Six   Bad News: The Indirect Approach   126

      Chapter Objectives   126
      Introduction   127
      Writing Bad News Messages   127
      Responding to Complaints or Problems   130
      Delivering Bad News Orally   139
      Chapter Review/Self-Test   140
      Exercises   142

Chapter Seven   Persuasive Messages   146

      Chapter Objectives   146
      Introduction   147

Elements of Persuasion   147
The AIDA Sequence   149
Sales Messages   151
Special Requests   158
Making a Complaint   161
Supervisory Communication   163
Persuasion in Oral Communication   167
Chapter Review/Self-Test   170
Exercises   172

Chapter Eight   Oral Communication within the Organization   175

Chapter Objectives   175
Introduction: The Importance of Oral Communication   176
Delivering Effective Oral Presentations   176
Meeting and Working in Small Groups   184
Technological Communication   191
Answers to Chapter Questions   193
Chapter Review/Self-Test   193
Exercises   195

PART THREE   The Job Search   199

Chapter Nine   Résumés   200

Chapter Objectives   200
Introduction   201
Pre-Résumé Analysis   201
Preparing a Résumé   208
Electronic Résumés   221
Chapter Review/Self-Test   223
Exercises   224

Chapter Ten   Locating and Applying for Jobs   226

Chapter Objectives   226
Introduction   227
Researching the Job Market   227
Writing an Unsolicited Letter of Application   229
Answering a Job Advertisement   234
Application Forms   239
Oral Communication: Replying to a Job Posting by Telephone   241

Chapter Review/Self-Test   243
Exercises   245

Chapter Eleven   Interviews   247

Chapter Objectives   247
Introduction   248
Preparing for the Interview   248
At the Interview   251
Follow-up to Interviews   253
Learning from the Interview   256
Standardized Tests   257
Chapter Review/Self-Test   259
Exercises   261

PART FOUR   **The Report   265**

Chapter Twelve   Researching the Report   266

Chapter Objectives   266
Introduction: What Is Research?   267
Being Sure of Your Purpose   267
Facts, Opinions, and Logical Conclusions   268
Primary and Secondary Research   274
Primary Research: Designing Questionnaires   275
Primary Research: Personal and Telephone Interviews   277
Secondary Research   281
Recording and Using Information from Sources   291
Documentation of Sources   293
Answers to Chapter Questions   298
Chapter Review/Self-Test   299
Exercises   301

Chapter Thirteen   Planning the Informal Report   304

Chapter Objectives   304
Introduction: What Is a Report?   305
Report Formats   305
Planning and Writing Informal Informational Reports   307
Printed Report Forms   314
Planning and Writing an Analytical Report   315
Persuasive Reports   319

Chapter Review/Self-Test 324
Exercises 326

Chapter Fourteen   Presenting the Formal Report   330

Chapter Objectives 330
Introduction: The Importance of Presentation 331
Structure of a Formal Report 331
Graphic Elements of Reports 350
Final Editing 355
Chapter Review/Self-Test 356
Exercises 358

APPENDIX   Handbook of Grammar and Usage   359

Introduction 360
Recognizable Patterns 360
Units in Grammar 361
Guidelines for Usage 365
Guidelines for Punctuation 381
Guidelines for Spelling, Handling Numbers, and Capitalizing 391

INDEX   401

Employability Skills Profile:
The Critical Skills Required of the Canadian Work Force

Academic Skills
Those skills which provide the basic foundation to get, keep, and progress on a job and to achieve the best results

Canadian employers need a person who can:

COMMUNICATE

- Understand and speak the languages in which business is conducted
- Listen to understand and learn
- Read, comprehend, and use written materials including graphs, charts, and displays
- Write effectively in the languages in which business is conducted

THINK

- Think critically and act logically to evaluate situations, solve problems, and make decisions
- Understand and solve problems involving mathematics and use the results
- Use technology, instruments, tools, and information systems effectively
- Access and apply specialized knowledge from various fields (e.g., skilled trades, technology, physical sciences, arts, and social sciences)

LEARN

- Continue to learn for life

Personal Management Skills
The combination of skills, attitudes, and behaviours required to get, keep, and progress on a job and to achieve the best results

Canadian employers need a person who can demonstrate:

POSITIVE ATTITUDES AND BEHAVIOURS

- Self-esteem and confidence
- Honesty, integrity, and personal ethics
- A positive attitude toward learning, growth, and personal health
- Initiative, energy, and persistence to get the job done

RESPONSIBILITY

- The ability to set goals and priorities in work and personal life
- The ability to plan and manage time, money, and other resources to achieve goals
- Accountability for actions taken

ADAPTABILITY

- A positive attitude toward change
- Recognition of and respect for people's diversity and individual differences
- The ability to identify and suggest new ideas to get the job done — creativity

Teamwork Skills
Those skills needed to work with others on a job and to achieve the best results

Canadian employers need a person who can:

WORK WITH OTHERS

- Understand and contribute to the organization's goals
- Understand and work within the culture of the group
- Plan and make decisions with others and support the outcomes
- Respect the thoughts and opinions of others in the group
- Exercise "give and take" to achieve group results
- Seek a team approach as appropriate
- Lead when appropriate, mobilizing the group for high performance

This profile outlines foundation skills for employability. For individuals and for schools, preparing for work or employability is one of several goals, all of which are important for society.

SOURCE: Employability Skills 2000+ Brochure 2000 E/F (Ottawa: The Conference Board of Canada 2000).

PART ONE

Introduction to Communication

CHAPTER ONE

THE COMMUNICATION PROCESS

Chapter Objectives

Successful completion of this chapter will enable you to:

1. Identify the three parts of the communication process.

2. Identify your purpose for communicating in a given situation.

3. Identify the needs and relevant characteristics of your audience.

4. Select the medium best suited to your purpose and audience.

5. Use nonverbal communication to enhance the effectiveness of a verbal message.

6. Plan your oral or written message to overcome communication barriers.

1 INTRODUCTION: THREE PARTS OF THE COMMUNICATION PROCESS

Communication is a process. A process is a set of actions or changes that brings about a specific result. In the communication process, a message is conveyed from sender to receiver. Because a message is something you can see or hear, it is tempting to think that the message *is* communication. But communication involves an interaction among sender, message, and receiver; it is a process, not a product. This point is important to remember when you are planning communication or trying to improve it.

In the communication process, the **sender** formulates an idea or set of ideas that he or she wishes to communicate to a receiver. The sender selects the verbal and/or nonverbal means that will be used to convey the idea to the receiver. In this way an idea becomes a **message.** The message is then transmitted to its audience by a channel or medium such as speech, writing, or gesture. The **receiver** receives the message—that is, he or she sees or hears it. The receiver must also **decode** the message, so that he or she understands the idea that the sender was attempting to convey (see also Figure 1.1).

FIGURE 1.1 The Communication Process

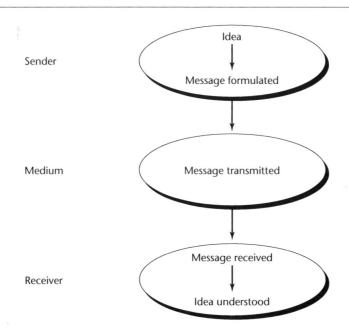

Planning Your Message

Getting all the elements of the communication process to interact successfully takes planning. Just as a well-run meeting has an agenda to make sure that important issues are discussed and decisions are made, a well-crafted message has a plan to ensure that the sender's ideas reach the receiver accurately and effectively. On page 13 you will see a sample **planning worksheet.** You wouldn't want to fill one out every time you had to write or speak, but you may find it a useful tool for planning your assignments in this course. By the end of the semester, the planning steps should have become internalized so that you automatically consider purpose, audience, medium, and message whenever you communicate.

2 IDENTIFYING YOUR PURPOSE

In our everyday social life, communication is often an end in itself. We enjoy sharing our ideas and feelings as a way of expressing ourselves and building relationships. At work, however, we are generally communicating to achieve a purpose related to the goals of the organization. When we say that the communication process has been a success, we mean that the sender's message has produced the desired impact on the audience. In your job, you may have little choice about messages that must be conveyed—your car is ready, you're overdrawn, your cheque is in the mail—but you are able to determine whether these messages are communicated effectively to your audience.

Try to express your purpose as something that can be seen or measured. This will help you evaluate the success of your message after you send it. For example, if you are sending an e-mail message to your supervisor about problems with your computer, think of your purpose as "getting a new computer" or "getting a visit from a technical support person" rather than as "letting my supervisor know I'm having problems." The more clearly you understand your purpose in communicating, the more effectively you can plan the format, the content, and the words that will help you achieve this purpose.

Remember that messages often have more than one purpose. If you are answering a customer's questions about your package tours, you are giving information, but you should also—if you're a good company representative—be trying to persuade the customer to choose your tour rather than someone else's. If you are turning down a request for a refund, you are conveying bad news, but you should also be trying to retain the goodwill of your audience toward your organization.

See if you can identify the purpose or purposes of the following messages:

(a)

BRING A FRIEND. BUY ONE, GET ONE FREE.

Present this card when you purchase a ticket for any of our shows and receive a second ticket with our compliments.

Lighthouse Theatres

Want people to fill seats

(b) Covering letter enclosed with a résumé.

grab their attention to look at resume

(c) Announcement of a special discount offer from a newly opened shop.

Get people in the door - need customers

(d)

want customers

(e) Letter to the manager of a hotel describing the unsatisfactory service you received.

looking for an apology, discount

(f) Tips on preventing car theft that are enclosed with insurance renewal notice.

be more careful + don't want to pay a policy

(g) Letter of thanks from a business to a printing company that rushed a special order.

stay in good standing - if need it again, I will help them

(h) Store plaque with name and picture of the "Employee of the Month."

respect hard working employee - motivate positive reinforcement

(i) Letter to an unsuccessful job applicant.

respect, encourage to keep trying down the road if something can up

(j)

> **AN URGENT MESSAGE TO OWNERS OF COFFEE CADDY COFFEE MAKERS: STOP USING THE COFFEE MAKER FOR ANY PURPOSE**
>
> - Some Coffee Caddy coffee makers can be dangerous to use.
> - The complete plastic top may separate from the glass pot without warning.
>
> If you have a Coffee Caddy coffee maker, please fill out and return the coupon below.
>
> Do not return the coffee maker. Keep it until you hear from Coffee Caddy. We will send you further information and a special offer.
>
> We regret any inconvenience; however, we are concerned for your safety. Thank you for your cooperation.

save law suits, public service announcement – care about customers

3 IDENTIFYING YOUR AUDIENCE

Once you have established your purpose in all its dimensions, your next consideration should be your **audience.** Who will be reading or listening to your message? In personal situations, you have already had extensive practice in shaping your message to your audience. For example, consider how you would tell a friend that you were behind on a project and how you would tell the same thing to a teacher from whom you were hoping to get an extension. A change of audience means a change of vocabulary, of tone, of content, and of length. The more you know, or can intuitively guess, about your audience, the better you can tailor your message to meet their needs.

Think about the experience and occupation of your audience. What general knowledge or special knowledge do they have about your subject? What is their attitude to your subject—friendly, hostile, neutral, or bored? What is your relationship to your audience? What direction will your message be travelling in? As you plan and/or write your message, imagine that you are getting feedback from your audience throughout the process. What are some of the questions your audience might ask you? Does your message answer typical questions like Why? How? and So what?

4 SELECTING THE APPROPRIATE MEDIUM

"The medium is the message." This statement by communications theorist Marshall McLuhan was a buzz phrase in the 1960s but is still relevant today. The means you use to transmit your message will contribute almost as much to its success in achieving its purpose as your choice of words will. Sometimes you will be told to "put it in writing." At other times, you will be asked to "say a few words" at a meeting or "get on the phone to head office." In these situations, you have no chance to decide which medium you would prefer. Sometimes, however, you will be able to choose whether to write, visit someone in person, phone, or use other technology. Your decision should be based on more than personal preference: the advantages of each medium should be carefully weighed.

Advantages of Oral Communication

1. **Speaking takes less time than writing.** The average person speaks at a rate of about 150 words a minute. Composing a 150-word message on paper, even a rough draft, would take much longer. Even the physical act of transcribing on a keyboard does not approach the speed of oral communication.
2. **There are fewer mechanical problems to worry about.** Grammar and punctuation are much more flexible in oral communication, and, of course, spelling doesn't exist in speech. Errors can be corrected immediately, and in any event they are less likely to be remembered, or even noticed, by the audience.
3. **The speaker's voice and body language add meaning and variety.** A speaker can communicate with gesture, tone, volume, eye contact, and physical appearance. These details not only expand and clarify the message but also hold the attention of the audience.
4. **Speaking allows feedback from the receiver.** Because the sender and receiver are together at the same time and often in the same place, it is easier for the sender to feel in control of the communication process and more confident that his or her message is being received. The more one-to-one feedback the situation permits, the more control the sender has, because an oral message can be easily revised to meet the needs of the receiver and thereby fulfill the sender's purpose.
5. **Speaking is more personal.** The audience generally feels closer to the sender and more involved in the communication process when a message is delivered orally. The speaker also generally feels closer to his or her audience. Speaking emphasizes the human element in communication.

Disadvantages of Oral Communication

Oral communication does, however, have its downside. Its special characteristics, so valuable in some situations, can pose a major barrier to communication in others.

1. **The speaker's voice, appearance, or body language can compete with the verbal message for the listener's attention.** Overlapping verbal and nonverbal messages often lead different listeners to come away from a meeting or a conversation with very different memories of what they "heard."
2. **Internal or external distractions can cause the listener to miss the message.** Your listeners may be more concerned with their personal problems or their plans for the weekend than they are with your message. Or they may be unable to hear or focus because of noise or activity going on at the same time.
3. **Speaking is more personal.** This factor becomes a disadvantage when conflict exists between speaker and listener. People often find it difficult to respond to the message rather than the messenger.
4. **Oral communication is most effective when speaker and listener are together at the same time.** It can be challenging to find a time when busy people are all available, especially if they are in different time zones.

Advantages of Written Communication

1. **The receiver must take an active role in the communication process.** Written communication places much more control in the hands of the receiver. Because a reader has so much control over the delivery of the message, a writer can cover more complex material than a speaker can and still be confident that the audience can take it all in.
2. **Reading is faster than listening.** You can read from three to five times more in an hour than you could hear delivered orally.
3. **Writing is less personal.** A written report makes it much easier to concentrate on the contents rather than on the author. Likewise, writers can distance themselves from negative feelings they may have about the intended audience. This distance also relieves the stress of anticipating some kind of immediate negative feedback.
4. **Writing provides a permanent record.** If you have ever forgotten someone's name five seconds after you were introduced, you will recognize the impossibility of retaining in your head even a tiny fraction of the information necessary to run a business. The economic and cultural structure of modern society depends on our ability to store and retrieve immense amounts of information.

Modern electronic media now make it possible to preserve oral records, but the expense, the trouble, and the difficulty of efficient access to oral records make written records better and more efficient.

5. **Writing can be revised.** It's the Big Meeting. You are scheduled to give a short report on your department's progress in employment equity. Just before your turn comes, the person beside you accidentally sets a jelly doughnut on your notes. You get to the podium, trying to be subtle as you wipe off your note cards. Unfortunately, the ink is coming off with the jelly. You're very nervous, so some of the connections you want to make get lost in the delivery. Things seem under control until you discover that your last two note cards are stuck together. You wind up as best you can and return to your seat. Only when you sit down do you notice the blob of jelly on your tie.

Well, you may console yourself that there will be other opportunities. But this one is gone for all time. Circumstances did not allow you to make the best of it. Writing, on the other hand, gives you an infinite number of second chances. If your first draft is a failure, there can be a second, third, fourth—until the final version represents the best effort you are capable of. Your audience will not be aware of the sweat and toil and the many inferior attempts; all they will see will be the polished final result.

Matching the Medium to Your Audience

Fortunately, modern technology can help you enjoy the advantages of speaking or writing in any given situation while overcoming the disadvantages. For example, e-mail has made it possible to send a message and receive feedback at high speed, while still allowing you to edit the message, send it to one person or many people, and keep a copy on file. Thus it combines some of the best features of the letter and the telephone call. Cell phones and voice mail have reduced the frustration of "telephone tag," where two people with conflicting office schedules could never reach one another by phone.

To select the medium that is best suited to your message, audience, and purpose, ask yourself how important each of the following criteria will be to you in this situation:

- **Speed.** How long can I wait for my message to be received?
- **Security.** Is privacy important? Is legal confidentiality an issue?
- **Visual impact.** Is it important for the message to look impressive?
- **Accessibility and permanence.** Do I expect my message to be referred to more than once?

- **Efficiency.** How important is it to me to reduce the cost or effort of sending this message?

Ask yourself about the needs of your audience as well. Generally everyone opens mail and answers the telephone. But many people will not open an e-mail message unless they know who sent it, and other people are slow to answer e-mails, or return phone messages, or do either. A letter, a phone call, or a phone message saying an e-mail is being sent are good ways to begin the communication process.

Generally it is best to respond to a message in the same medium in which it was sent. The person who left a voice-mail message will be expecting a call back; the person who e-mailed you will be checking his or her messages for your reply.

Suppose you wished to transmit some detailed tax calculations to your supervisor. Should you deliver them orally by phone, or in person? Spoken communication is faster, more interesting and personal, and allows feedback from the receiver. How relevant are these considerations to your task? Written communication allows the receiver to control communication, takes less time on the part of the receiver, and provides a permanent record. Clearly, these considerations far outweigh the advantages of oral communication in this case. Writing allows busy administrators to receive the information at their convenience. They can spend as much time as necessary to go through the complex data and then file the information for later reference. The transmission of written material within an organization is usually rapid.

In another situation, a client might have called with a problem. You need to negotiate a solution. Speaking gives you the immediate two-way communication you require. Or, imagine that you are soliciting donations to the Canadian Coalition Against Acne. Because the sender has greater control in speaking than in writing, you choose a phone campaign, knowing that it is harder to hang up on someone than it is to sling junk mail in the wastebasket. Or you go door to door, because personal contact makes it yet more difficult to refuse a request. Often a stressful situation, such as handling a complaint or firing someone, requires all the resources of face-to-face discussion.

Here are some examples of communication situations in which you might find yourself. Assume that you have access to all of the communication media on the list below. For each of the following situations, write the number of the medium you would use on the blank line. In case of a tie, put both numbers, joined by "or." If you would use more than one medium in a situation, put both numbers, joined by "and."

1. personal visit
2. phone call
3. voice mail
4. letter
5. fax
6. paper memo
7. e-mail

_2_ 1. You need to confirm the location of a meeting taking place tomorrow.
_1_ 2. A colleague who works at another office has sent you a draft of the proposal you are working on together. You find some errors when you are proofreading it. Your colleague has the disk.
_2_ 3. A long-time client in another city has written an angry letter about a major foul-up.
_4_ 4. You wish to request a leave of absence from your job.
_7, 4_ 5. You are coordinating a national conference. Since the time you sent out the information package to registrants, there have been important changes to the program. The conference begins in two weeks.
_4, 6, 7_ 6. An employee has asked you for details about company life insurance benefits.
_1_ 7. You want to introduce yourself to a new employee.
_2, 7_ 8. Your former supervisor, who transferred to head office six months ago, has been promoted to vice-president. You want to congratulate her.
_1, 6_ 9. Another employee has been helping herself to goodies you've stashed in the fridge.
_7, 2_ 10. A shipment of gazingus wires arrived from Vancouver minus 52 units. You must inform the supplier.

5 NONVERBAL MESSAGES

Nonverbal messages can compete with verbal messages and create distraction. In these situations, nonverbal messages create barriers to effective communication. Ideally, however, nonverbal messages such as appearance, gesture, expression, and tone of voice should support what you write or say and increase its positive effect on your audience. For example, a crisply printed report with a well-laid-out title page and an appropriate cover will suggest that the writer has put thought and effort into the content as well. The following sections discuss some basic ways to enhance communication nonverbally.

PLANNING WORKSHEET

Purpose
Why am I sending this message? What will be the measurable result if my message is successful?

_____
_____
_____
_____

Audience
Who will be reading or listening? What do I know, or what can I predict about him/her/them?

_____
_____
_____
_____

Medium
How will I be sending this message? Why is this the best choice?

_____
_____
_____
_____

Message
What information does my audience want/need to achieve my purpose?

_____
_____
_____
_____

Appearance and Grooming

If you hope to be a supervisor some day, observe what people in senior jobs wear and how they look. It's more important to reflect the standards of your workplace than to make a fashion statement. If your job as a salesperson, for example, takes you to many different kinds of workplaces, a conservative, tailored look is a safe choice. Invest in a good briefcase, and clean it out and organize it regularly. Never root around in it in front of supervisors or clients. Buy an expensive, or at least expensive-looking, pen.

Gesture and Body Language

Whether you are speaking or listening, your body language should show that you are involved with the communication process. Lolling back in your chair, folding your arms, looking away or carrying out tasks like straightening your desk, all send the opposite message. Maintain eye contact during conversation and turn to face the speaker in a meeting. If you are sitting, avoid slumping, but don't sit too rigidly; it's normal to listen actively with your whole body. Keep your arms off the arms of the chair. When you are speaking, it is natural to emphasize your point with your hands; some people do this more than others. Just avoid gestures that can be seen as hostile, such as touching or finger-pointing. Negative body language, such as eye-rolling, pinching the lips, or heaving a sigh, is just as offensive as a negative comment.

Tone of Voice

The human voice is a very expressive instrument, and with a simple change of emphasis, anyone can turn a statement like "How great to see you!" from a friendly greeting to a crushing insult. Even a voice without expression communicates a nonverbal message; when a telemarketer says, "I'm sure you, like all Canadians, are very interested in receiving liposuction in the privacy of your home," in a complete monotone, we know that he or she is reading off a script. More and more people in service jobs, in both the public and private sectors, are also being trained to use a kind of script; for example, to greet every caller by saying "I'm X. How may I help you today?" In theory, this should get across the message that the agency or the business is here to help you, the client. In practice, depending on the way in which the greeting is spoken, the message is often, "Why are you bothering me?"

The fact is that if there is a choice between believing the words or the tone of voice, your listener will choose the tone of voice every time.

6 ELIMINATING COMMUNICATION BARRIERS

Earlier in this chapter you were encouraged to express the purpose of a message as something observable or measurable, so you would be able to judge whether or not you had communicated successfully. What happens if your purpose is not achieved? How can you do better next time? The planning worksheet is a useful tool for finding barriers to communication and creating a better message when the first attempt has been unsuccessful. Look at your original plan, or fill out the sheet if you did not fill it out before. Then ask yourself the following questions:

Purpose
Is my purpose clear to me? If I have more than one purpose, do they conflict in any way? Do I have a way of measuring the success of my message?

Audience
Do I know the needs of my audience? Have I tried to put myself in the reader's or listener's place in creating this message to make it clear, complete, and respectful?

Medium
Did I decide to send the message this way because it was convenient and familiar, or did I choose the best medium for my audience and message? Did unexpected events prevent delivery of my message? What backup plan do I need in case of technical failure or human mistakes?

Message
Is everything in my message relevant to my audience and purpose? Am I sending a nonverbal message that conflicts with my purpose?

CHAPTER REVIEW/SELF-TEST

1. "Communication" is
    (a) an oral or written message.
    (b) a three-part process.
    (c) a medium for transmitting ideas.
    (d) all of the above.

2. The first step in communication planning is
   (a) choosing the appropriate format.
   (b) discovering and expressing your feelings on the subject.
   (c) building a relationship with the receiver.
   (d) identifying the measurable result if your message is successful.

3. Knowing the needs and characteristics of the intended receiver of your message will help you choose
   (a) the length of the message.
   (b) the vocabulary and sentence structure of the message.
   (c) the tone of the message.
   (d) all of the above.

4. If you have the choice of how to send your message, you should consider
   (a) what is fastest and most convenient for you.
   (b) which medium will be most likely to help you achieve your purpose.
   (c) which medium is the most commonly used.
   (d) your personal preferences.

5. Oral communication is a good choice if
   (a) you are delivering bad news to a coworker.
   (b) your message is long and detailed.
   (c) your relationship with the receiver has been negative.
   (d) your receiver is very busy.

6. Written communication is a good choice if
   (a) your message is urgent.
   (b) you haven't had a chance to think much about the subject.
   (c) your message has legal importance.
   (d) your receiver may consider your subject boring or unpleasant.

7. Written communication is particularly important in the workplace because
   (a) it can be revised for accuracy and good style.
   (b) it provides a permanent record.
   (c) it is easy to store and retrieve.
   (d) all of the above.

8. E-mail can be used in any situation where written communication is appropriate except
   (a) with elderly clients or customers.
   (b) when the contents are confidential.
   (c) when it is important that the message have no spelling or grammar errors.
   (d) when the message must be well laid out on the page.

9. Use electronic communication to
   (a) combine the advantages of oral and written communication.
   (b) impress clients and coworkers with your state-of-the-art equipment.
   (c) compensate for weaknesses in your message.
   (d) all of the above.

10. When your message does not achieve its intended purpose, you should
    (a) blame the receiver.
    (b) send it again.
    (c) use a planning worksheet to analyze your audience and identify possible problems with your purpose, medium, or message.
    (d) upgrade your technology.

Answers

1. B 2. D 3. D 4. B 5. A 6. C 7. D 8. B 9. A 10. C

EXERCISES

1. John Green had been running the marketing department at World Trek Tours for 12 years. Perhaps "running" wasn't quite the word; John's attitude was that since the employees all knew their jobs, he should just let them get on with it. As a result, the department was inefficient but friendly. The employees were genuinely shocked when John got the axe, apparently without warning, just before it was time to start planning the summer holiday promotions. No one knew anything about his replacement, Ed Harley. According to the grapevine, Ed was right out of a marketing program, with no experience in the travel industry. Someone had suggested that his primary purpose was to shake up the department and send some people out the door after John Green. So Dan wasn't exactly looking forward to his first meeting with Ed.

   As head of the advertising department for eight years and a personal friend of John Green, Dan was closely identified with the "old guard." Ed had scheduled the meeting for 9:30; by 9:50, sitting in the reception area, Dan had had plenty of time to become nervous. Through the door he could hear Ed's voice: "Well, I'm not talking about soya beans, sweetheart, so what are you telling me? See if you can get your rear in gear." The phone slammed down, and Ed's door burst open. "Dan—c'mon in and sit down." Dan was motioned to an armless swivel chair. Ed continued to stand, staring out the window, with his back to Dan. Suddenly he swung around. "Dan, can you answer one question for me? Why is the marketing department such a complete and utter disaster area?" Dan's eyes goggled as he took this in.

"Now, I know," Ed continued, "that John Green was a burnt-out incompetent who didn't know a marketing mix from a cake mix, and I know that this department has been a refuge for every loser and lead-swinger in the company, but I'm still mystified, genuinely mystified, at how *totally* inept the whole operation has become. Do you recognize this?" Ed whipped a sheet of paper off his desk and held it an inch from Dan's nose. "No, I'm ... that is ... " Dan tried to focus on the page. "It's a memo regarding planning the summer promotions," Ed answered his own question. "Oh, yeah—Tina James and Ted Burton put that out. It's ... " "It's a hunk of crud," Ed snapped. "I wouldn't use these guidelines to plan a trip to the men's room, let alone a $1.3 million campaign! What's going on here?"

"Well, I'm ... " Dan cleared his throat. "Well, I'm not in on that part of the process. They usually get that stuff from Brenda and Roy, and then the research team works on it. I'm not involved until ... " "I don't care if they got it from Mickey and Goofy. It's not worth the paper to print it." Ed moved over behind Dan's chair. "Now, things here have just—bumbled along; everybody doing his own thing and all one big happy family. Well, let's get one thing straight." Ed swivelled the chair around and bent down until he and Dan were nose-to-nose. "The party's over." He straightened up. "That's all I have to say. You can go now." Dan made his way toward the door. Ed picked up the phone and started to dial. "Close the door behind you, would you."

Use the following questions to identify some barriers to communication in this situation:
(a) What communication barriers existed between Ed and Dan even before they met?
(b) What nonverbal messages were sent from Ed to Dan? How would these messages be likely to affect Dan?
(c) What statements by Ed would be likely to create communication barriers between him and Dan?
(d) What impression of Ed will Dan convey to his fellow workers? Will this help Ed communicate effectively in future interviews with other personnel?
(e) Ed seems to be using intimidation as a management strategy. How can this backfire?

2. Identify barriers to communication that could arise in the following situations:
   (a) A caller begins a telephone conversation with the comment, "You're a hard person to get hold of."
   (b) During a business meeting, your cell phone rings several times. Your ring tone is a version of "The Drugs Aren't Working."
   (c) A job hunter uses her present employer's machine to fax résumés to other companies.
   (d) A company produces a brochure to recruit college graduates. Pictures of a diverse workplace appear to have been created by Photoshop.

(e) A job hunter applies a generous splash of Olde Gymbagg aftershave before he leaves for an interview.
(f) A customer phones and is kept on hold for 17 minutes. Every 30 seconds she is told, "Your call is important to us."
(g) A fundraising auction for the Animal Welfare League includes a fur stole and a barbeque.

3. Select an example of any of the following: magazine advertisement, office memo, instruction manual or package insert, informational or promotional brochure, personal letter, short business report, "junk" mail. In a brief essay (250 words):
   (a) describe why the message was delivered in writing rather than orally;
   (b) state what changes would have to be made in the text of the message to adapt it to effective oral communication.

4. You are the owner/manager of Head-to-Toe Beauty Spa. Decide whether you would handle the following communication situations in person (P), in writing (W), or by telephone (T), or write in another technology, such as fax or e-mail:
   (a) You want to contact the placement service of your local community college to recruit students for part-time jobs. _____
   (b) You want to contact a European supplier about a product line that you recently read about. _____
   (c) You want to let your regular customers know about upcoming specials. _____
   (d) You discovered some items missing in a recent shipment of beauty supplies. _____
   (e) Customers have complained that your massage therapist has cold hands; you want to let her know. _____
   (f) You need information about municipal health regulations. _____
   (g) You want to tell someone who recently sent you a résumé that you have no full-time openings. _____
   (h) Because of a cancellation, an appointment is available tomorrow for a waiting customer. _____
   (i) A cheque from a regular customer was recently returned NSF from the bank. _____
   (j) You need some advice about expanding the floor area of your shop. _____

5. Prepare a **planning worksheet,** following the example on page 13, for one of the following situations:
   (a) Your daughter's soccer team needs a sponsor. You wish to approach a local business for support.
   (b) A friend is looking for a job. You want to get her an interview at your workplace.

(c) A client regularly sends you e-mail, which your workplace computer thinks is "spam." You would like him to use a different server to avoid this problem, which has been going on for several months.

(d) Your workplace currently rewards the "Employee of the Month" with a special parking spot near the front door. You would like to see an appropriate alternative offered to any "Employee of the Month" who doesn't drive to work.

(e) You have evidence that one of your employees is using his workplace computer to sell items on the Internet during working hours.

CHAPTER TWO

STYLE IN BUSINESS COMMUNICATION

Chapter Objectives

Successful completion of this chapter will enable you to:

1. Edit your written messages to eliminate wordiness and unclear phrasing.

2. Deliver a clear, concise oral message.

3. Edit your written communication to ensure a positive, reader-oriented, and non-discriminatory tone.

4. Use a positive, audience-oriented, and non-discriminatory tone in your oral communication.

INTRODUCTION

You may think at first that the word "style" is out of place in a book about communicating in the workplace. It may suggest to you a kind of literary decoration, a polished surface that has nothing to do with the real content of the message. It is true that on the job we are concerned with getting a task done, not impressing others with our artistic use of language. But while the style of a poem or novel might be inappropriate in the workplace, everything you say or write has a style. You want to find the style that is most effective in achieving the purpose of your message.

Some people still think that "business writing" requires special jargon—long words for simple ideas, phrases like "as per your request" or "with regard to the above." In fact, these are leftovers from a more formal era, no more appropriate in today's environment than green eyeshades or quill pens. Contemporary practice is simply to adapt good English to work situations by applying a few simple principles:

1. Make your message clear by being concise and using simple, concrete words.
2. Proofread to ensure standard spelling, grammar, and punctuation.
3. Emphasize the importance of your audience by using the words "you" and "your" as often as possible.
4. Create a positive impression by avoiding words with strong negative overtones, putting statements in positive form, and emphasizing positive characteristics.
5. Eliminate racist or sexist bias in speaking and writing.

1 WRITING FOR CLARITY

Unclear communication causes costly delays and mistakes. Your style must be clear to ensure that your audience receives your ideas quickly and accurately.

CASE 1

Brad had worked hard on his message for Mr. Santos. He knew that communication skills were really important in the department, and he wanted Mr. Santos to know that he could be counted on to carry some of the responsibility for report writing and correspondence. So this assignment had been a big challenge. Mr. Santos had just picked up Brad's draft from his mailbox. If he liked it, the whole department would soon be reading it. Of course, it would be under Mr. Santos's name, so Brad had tried to make it sound as businesslike and formal as he could. "Those of us who are on the management team have identified the personnel situation as an area of concern. Something that could be done about this is to implement regular opportunities for group discussion, which could take place on a scheduled basis without interfering with your normal work routines. If you are of the opinion that this could change things in a positive direction, please come prepared at a meeting which will be held in the boardroom next Wednesday morning at 10 AM."

About an hour later Brad found a message in his mailbox:

"Subject: Meeting, Wednesday 10 AM"

Brad started to read eagerly. But, wait a minute — almost every word had been changed. Mr. Santos didn't like his draft. What had gone wrong?

Why had Mr. Santos not used Brad's version of the message? Unfortunately for Brad, his writing lacked conciseness and clarity. His message was obscured by vague, wordy phrases, while Mr. Santos wanted something that could be read and understood quickly and precisely. Here is Mr. Santos's version:

Are you concerned about staff cutbacks? If you are, come to a meeting and share your views. Several meeting times will be offered to fit your work schedule. The first will be held this Wednesday at 10 AM in the boardroom.

Here are some suggestions to help you avoid Brad's mistakes and achieve a clear, concise, and readable style.

Conciseness

Planned repetition—for example, repeating an important deadline at the end of a memo—can help your audience understand and remember your message better. Unplanned repetition—redundancy—wastes reading and listening time and may confuse your audience. Eliminate redundant expressions like the following:

Redundant	**Concise**
have need of	need
for the purpose of	for
the amount of $38.50	$38.50
period of time	time
red in colour	red
the city of Edmonton	Edmonton
attractive in appearance	attractive
the month of December	December
I would like to take this occasion to …	I would like to …

that – very overused

past experience	experience
until such time as	until
he is engaged in researching	he is researching
refer back	refer
a matter of managing	managing
at a time when	when
on the subject of	on

Some expressions, while not redundant, contain more words than are needed to express the thought. While it may not seem worth your while to eliminate one or two words, think of it in percentage terms. Reducing a ten-word sentence to eight words is a 20 percent saving; it means reducing a five-page report to four pages, with the resulting saving in keyboarding, paper, photocopying, and so on. Here are some examples of wordy expressions. Write a concise version beside each.

Wordy	**Concise**
due to the fact that	due to or because
in the neighbourhood of	around / about / near
in the event that	in case / if
it may be that	possibly, perhaps
at an early date	soon
in reference to	about
at this point in time	now
sometime in the future	later
it would appear	apparently
in view of the fact that	since, because

Do not feel that business situations demand special long-winded phrases. Contemporary practice recognizes that the appropriate style for business is clear English. Expressions such as the following should be eliminated altogether, since they add nothing to the message:

it has come to my attention
this is to inform you
in closing

please be advised that
I am writing to tell you that

Examples

✗ POOR
It has come to my attention that not everyone is contributing to the coffee money.

✓ IMPROVED
Not everyone is contributing to the coffee money.

✗ POOR
I am writing to inform you that your application for our management trainee program has been received.

✓ IMPROVED
Thank you for applying to our management trainee program.

Some old-fashioned expressions do carry a message, although the vocabulary is so outdated that the meaning becomes obscured. You can express the same ideas in simple vocabulary that is part of your everyday speech.

Examples

✗ POOR
Enclosed herewith please find a map of Winnipeg as per your request.

✓ IMPROVED
Here is the map of Winnipeg you asked for.

✗ POOR
Kindly respond at your earliest convenience.

✓ IMPROVED
Please let us know as soon as possible.

Simple Vocabulary

As a general rule, choose the simplest, commonest words that express your meaning. When you use familiar words that are part of your everyday vocabulary, you will have more control over your message. You will avoid looking phony or pretentious, or saying something you didn't intend.

Furthermore, your message is less likely to be misinterpreted when you use simple, ordinary words, regardless of the educational or job status of the audience. Do not think that the workplace demands fancy, formal language. A clear, readable message will always sound professional. Though the jargon specific to your job can help make a message more concise and exact if your audience is familiar with the terms you are using, it is not appropriate for an audience unfamiliar with your job. Jargon borrowed from other disciplines quickly becomes stale and clichéd—for example: parameters, equation, reference or impact (used as verbs), bottom line, final analysis, touch base, materialize, and many others.

Find simpler, more familiar substitutes for the following words:

Unfamiliar	Simple
ascertain	to get, to find out
terminate	end
endeavour	try
commence	begin, start
indicate	show
initiate	begin, start
remuneration	payment
pursuant	about
subsequent	later
facilitate	help

ESL TIP

Communicating in a second or third language is particularly challenging in the workplace. It is not enough for your message to be understandable; it must also be "correct." And it is not even enough for your message to be correct; it must also be "idiomatic." That is, it must be expressed in a way that is typical of a native English speaker. A very formal style will sound strange and old-fashioned in today's workplace, even if your dictionary or writer's handbook tells you that it is grammatically correct. On the other hand, many expressions that native English speakers use in everyday speech are too casual for the office, especially written communication. Sentences such as "Our sales presentation bombed," "The auditors ratted us out," or "The legal department messed up big time," would never be acceptable in a report or an e-mail going to file. Dictionaries and grammar checks on your computer are not very helpful here as these expressions may be acceptable standard English when used in a different context. You need to train your own eye and ear by reading good examples of workplace writing. Remember that books, newspapers, and magazines are written to entertain as well as inform. Your goals in communication at work are much narrower. Try reading annual reports; customer information brochures given out at banks, cable and phone companies; product user's manuals. These are generally written by professionals in the corporate communication field and put into practice all the objectives of this course.

Concrete Words

A concrete word is one that denotes something that exists in the material world. The opposite, an abstract word, is one that denotes a concept or quality. "Many people are out of work" is a concrete sentence; "Unemployment is high" is an abstract one. Concrete words and sentences are appropriate in the workplace because they emphasize the fact that actions carried out by people cause things to happen. Abstract language often gives the impression that situations are unchanging and unchangeable because that's the way the world is. So we have:

The availability of funding is diminishing.

instead of:

Money is getting scarcer.

or:

Cessation of buying has set in.

rather than:

People have stopped buying.

Some writers like to use vague, abstract words to disguise the fact that they have only vague, unformed ideas about their subject. Unfortunately, these words tend to cloud their ideas even further. Avoid these abstractions:

situation
aspect
position
concept
basis
factor

Examples

✗ POOR
The development of the staffing situation is proceeding quite well.

✓ IMPROVED
Three qualified caseworkers have joined our staff.

✗ POOR
The position will soon be reached where all vacancies will have been filled.

✓ IMPROVED
Soon all vacancies will have been filled.

✗ POOR
This situation is the basis for our position on the concept of expansion.

✓ IMPROVED
Because of a steady increase in orders, we plan to expand.

Make an effort to eliminate sentences beginning with "There is/are ..." and "It is...."

Examples

✗ POOR
There are a number of reasons why people come in late.

✓ IMPROVED
People come in late for a number of reasons.

✗ POOR
It is a long drive from Winnipeg to Toronto.

✓ IMPROVED
Winnipeg is a long drive from Toronto.

Standard Spelling, Grammar, and Punctuation

A letter, e-mail, or report that contains mechanical errors will leave a poor impression of the writer's competence. It may also leave the wrong impression of the writer's message. Standard spelling, punctuation, and sentence structure are closely related to meaning and thus are a vital part of communication. Without punctuation, a sentence like "Alex loved Laura and Kathy and Vicki loved him" is very hard to decipher. Nonstandard punctuation puts up barriers to the reader's understanding. For example, "The first movie, made by David Cronenberg, was *Transfer*" implies that David Cronenberg invented the movie.

Likewise, faulty sentence structure leaves doubt in the mind of your reader. "Lying on a shelf, I spotted the book I was looking for" literally means that the speaker was lying on a shelf. Common sense and sentence structure are in conflict. The reader can never be completely certain what is meant. Spelling errors that change meaning usually involve proper names, but problems can also arise when similar-sounding or -looking words are confused. For example, "We didn't mean to except Larry" means "We didn't mean to leave Larry out," whereas "We didn't mean to accept Larry" means almost the opposite.

Here are some statements whose meanings are obscured by nonstandard grammar or punctuation. Rewrite each sentence to make its meaning clear.

1. Never give an apple to a baby that hasn't been peeled.

   Never give an unpeeled apple to a baby.

2. Having corrected the errors, the contract was ready to be signed by the agent.

   The contract was corrected and ready for the agent's signature.

3. My supervisor finally read the report I had written on the weekend.

   My supervisor read my my report on the weekend.

4. Rowing lifting weights running all of these can improve cardiovascular fitness.

   Rowing, lifting, weight, and running can improve cardiovascular fitness.

5. The Health and Safety Committee held a lunch-hour forum on safe sex in the employee lounge.

   The H and S Com held a lunch-hour forum in the employee lounge about safe sex.

The handbook at the end of this text will help you with questions about punctuation and sentence structure in your own writing.

2 CLARITY IN ORAL COMMUNICATION

Conciseness

Although people listen more slowly than they read, their ability to absorb detailed information through listening is not very great. Consequently, oral communication needs much more repetition to ensure that the message gets across. The secret is knowing when enough turns into too much. Obvious repetition will bore and frustrate your audience. In addition, wordy phrases will slow down the pace at which real information is delivered and tempt your listeners to let their attention wander.

Word Choice

Whatever you have learned about choosing simple, concrete words in writing is even more important in speaking. No one is going to get up and consult a dictionary while you are talking. Your message cannot be examined once it's delivered, so your meaning must be immediately evident to your listeners.

Speaking offers many communication channels, such as words, gestures, and facial expressions. Give them all time to do their job; don't overwhelm your listeners with words.

Pronunciation

Standard pronunciation contributes to clarity in speaking in the same way that standard spelling contributes to clear writing. Standard pronunciation ensures that your audience correctly identifies the word you are trying to use. Here are some guidelines to help you avoid errors:

1. **Confirm the pronunciation of any proper names you plan to use.** For example, if you will be mentioning fellow workers in a meeting or presentation, it is easy to check the preferred pronunciation of their names with them ahead of time. Phone the office of a prospective employer or client with whom you will be meeting, to check with a secretary or receptionist. Many people are very sensitive about their names, so your effort will be appreciated.

2. **If you discover a new word in print, check it in a dictionary** or with a trustworthy colleague before you try it out in public. The English language is full of traps for the unwary; the pronunciation of words such as "epitome," "misled," and "rationale" is not obvious.

3. **Listen for lazy pronunciation habits.** Saying things like "akkrit" for "accurate," "could of" for "could have," or "innawinna" for "in the window" blurs meaning and detracts from your professional image.

3 TONE

Tone is the subjective element of style, the overall feeling or impression conveyed to the audience by the sender's choice of words.

To a computer, clarity is everything. If it can understand a command, it will carry it out. A computer never gets in a snit, changes its mind, or loses interest. But human beings are not computers. The command "Sit down and shut up" is admirably clear and concise, yet it could not be used to call a meeting to order without creating deep offence. Human beings respond as much or more to the tone of a message as to its content. Your experience as a communicator has taught you a lot about choosing words that will make the tone of your message appropriate to your audience, and of course you will continue to apply these skills when you are a full-time worker. There are, however, a few elements of tone that are particularly important on the job, although you may not have given them much thought in everyday life.

You-Centred Tone

First of all, communication on the job demonstrates the importance of the audience by using the words "you" and "your" as much as possible. For most of us, what we do at work involves the active cooperation of other people. They must approve a recommendation, or carry out instructions, or buy a product, or respond in some other way to what we say and do. Communication at work underlines this cooperative element by making the audience the subject of what is said or written as often as possible. The simplest application of this is replacing "we" and "I" with "you."

Examples

✗ POOR
We are attaching some pictures of our new conference facilities.

✓ IMPROVED
You will find attached some pictures of the new conference facilities available for your group.

✗ POOR
We are open Thursday and Friday nights until nine.

✓ IMPROVED
You can use the drop-in centre until 9 PM on Thursday and Friday nights.

The word "you" attracts the reader or listener and tells him or her that the message is relevant and requires a response. (This response may be simply paying attention.) On the other hand, "I" and "my," or "we" and "our," send the message that the speaker or writer is chiefly interested in himself or herself. An employee reading a memo in which every sentence begins with "we"—"We have planned ... ," "We have decided ... ," "We have introduced ... "—isn't likely to feel that his or her concerns or suggestions have much value. A potential employer listening to a candidate who spends the whole interview saying, "I'm looking for ... ," "I'm interested in ... ," "I want to ... ," will assume that this candidate sees the job as a mere stepping stone in his or her career path.

Rewrite the following passages to make them more you-centred:

(a) To: All Employees
Starting January 15, all employees are asked to park in the north parking lot only.

Starting Jan 15, please park in the north parking lot.

(b) We offer a full line of accounting and tax services for small businesses.

Your small business could benefit from our full line of accounting and tax services.

(c) For the convenience of our customers, we are introducing our new phone order service.

For your convenience, a new phone order service has been introduced.

(d) We would be pleased to have the opportunity to answer your questions about our new product.

You can contact us with any questions on our new product.

(e) I hope you will be able to come to our meeting, as I am not very familiar with the issues and I'll need some support.

Your input and support would be very helpful at our meeting.

Of course, this can't be just a mechanical exercise in substituting "you" for "I" or "me," although it may start out that way. To sound genuinely "you-centred," put yourself in your audience's place and see things from their perspective. Doing this will improve not only the tone of your writing or speaking, but your content and message as well.

CASE 2

Maria looked at the pile of applications on her desk with a sinking heart. The personnel department staff had screened them all for the minimum qualifications, but there were still at least 50 qualified applicants to be considered. How could she get them down to a short list of ten in time for the first interview appointment? Reluctantly, she picked up the first covering letter. "Dear Ms. Palma:" it began, "I would like to apply for the job of General Accountant which I saw advertised in this morning's *Daily Times-Bugle*. I will be graduating from Pacific College this June with a diploma in Accounting and Finance, and a job with a moderate-sized firm like Mountain View Tours would give me the practical experience I need to choose a field for further specialization. I am enclosing a copy of my résumé with details of my education and experience. I hope I will be hearing from you soon, as it will make the semester a lot more relaxed if I know I have a job lined up. Sincerely, Lotta Ego." Maria watched the letter flutter gracefully into the wastebasket. A few more like this and she would be down to a short list in no time.

Every sentence in Lotta's letter began with the word "I." In fact, she used the word "I", "me," or "my" twelve times, and the word "you" only once. But this is only a symptom of her deeper lack of concern for the needs of her audience. Lotta described what this job had to offer her, not what she could offer Mountain View Tours. Telling her reader about her plans to "move on" was just another way of saying, "Invest time and money training me, and then watch me take my newly acquired skills elsewhere." In addition, she implied that she was destined for better things than Mountain View Tours—the company for which her reader worked.

Here is Lotta's letter rewritten in "you-centred" style:

> Dear Ms. Palma:
>
> I am applying for the job of General Accountant, which you advertised in the April 12 *Daily Times-Bugle*. As you can see from the attached résumé, I will be graduating from Pacific College this June with a diploma in Accounting and Finance. Your firm would benefit from my thorough grounding in accounting practice and previous experience in the travel industry. I would appreciate an opportunity to discuss this position with you personally.

This version uses the words "I" and "my" four times and the word "you" or "your" four times. More important, it stresses what Lotta has to offer Mountain View Tours, not what it has to offer her. Lotta's first letter said, "Do me a favour, give me a job." This one says, "Do *yourself* a favour."

Of course, the purpose of emphasizing "you" and "your" is to put the reader or listener in the picture. If your message implies blame or criticism, it would be more tactful to leave out "you" and "your," perhaps by using the passive voice as in the following examples:

✗ POOR
You must return both copies of the form next time.

✓ IMPROVED
Both copies of the form must be returned next time.

✗ POOR
You did not complete the paperwork for the GST refund.

✓ IMPROVED
The paperwork for the GST refund was not completed.

Positive Tone

Just as the repetition of "we" or "I" throughout a message creates an overall impression of self-centredness and lack of concern for the reader, the repetition of negative words and phrases can create a negative impression that goes beyond the content of the message. A positive tone extends the sense of cooperation that is created by you-centred communication. It emphasizes what you or your organization **can** do, not what you can't; what is possible, rather than what is impossible. This reinforces values like negotiation and problem-solving that are vital to a healthy working environment. Here are four ways to create a positive tone in your communication:

1. **Avoid words with strong negative overtones, such as**

deny	unacceptable
reject	incompetent
refuse	ignorant
regret	neglect
impossible	complaint
fail	fault

These are powerful words that convey feelings of judgment and even punishment, and they make most people uncomfortable. Attention will be focused on these feelings instead of on your message. These feelings are intensified if the word "you" is used—for example, "Your suggestion has been rejected," "Your question shows that you are ignorant of our return policy," or "You have failed to follow instructions." Do not risk clouding the issue by using these loaded words or similar expressions. Find more objective, emotionally neutral phrases.

Examples

✗ POOR
Your complaint has been forwarded to the payroll office.

✓ IMPROVED
Your concerns have been forwarded to the payroll office.

✗ POOR
You failed the aptitude test.

✓ IMPROVED
Your mark on the aptitude test was below 65.

Write improved versions of the following sentences. Compare your answers with the ones on page 46.

✗ POOR
(a) You have neglected to turn off the photocopier.

✓ IMPROVED
The photocopier was left on.

✗ POOR
(b) It is impossible to fill your order as requested.

✓ IMPROVED
We would be happy to fill this portion of your order as soon as possible.

Always give something positive

✗ POOR
(c) The committee found your solution unacceptable.

✓ IMPROVED
The committee decided on this solution:

2. **Do not apologize for your message unless it is clearly negative.** Words and phrases like "unfortunately" and "I regret to inform you" tell your audience that anger or disappointment are appropriate responses to your message. Use them only if the circumstances are serious and an apology or expression of sympathy is appropriate—for example, "Unfortunately, the theft of these items is not covered by your insurance policy." Do not turn a neutral message into bad news by using negative expressions.

   Examples

   ✗ POOR
   Unfortunately, we'll be meeting in a different room next week.

   ✓ IMPROVED
   We'll be meeting in a different room next week.

3. **Put statements in positive form.** Look at a sentence such as "Your interview cannot be scheduled before 3 o'clock." Regardless of when you would prefer to be interviewed, the tone of the sentence suggests that you are being deprived of something. Change the sentence to "Your interview can be scheduled any time after three o'clock," and that sense of deprivation disappears.

   Examples

   ✗ POOR
   Your account cannot be credited without a transaction number.

   ✓ IMPROVED
   As soon as you give us the transaction number, your account can be credited.

   ✗ POOR
   Local bus service to the site is only available during July and August. The rest of the year, you have to get there by car.

   ✓ IMPROVED
   You can get to the site year-round by car, or by local bus service during July and August.

Edit the following sentences to emphasize the positive:

(a) The conference room will not be vacant until noon.

The conference room will be available at noon.

(b) We do not provide a pool insurance policy; however, your pool is already covered by your house insurance.

We are pleased to inform you your house insurance includes coverage of your pool.

(c) Your stationery order cannot be filled until Thursday.

We would be pleased to fill your order on Thursday.

(d) Pre-authorized payment will ensure that your bills are ~~never paid late,~~ *always paid on time* even when you're away.

(e) This item is available only in white, grey, or tan.

This item is available in white, grey or tan.

Compare your answers with those on page 46.

4. **Create a positive association for ideas or products you are promoting.** When promoting an idea or a product, tell your audience about the good features it has, not the bad ones it doesn't have. Look at this example: "Beautiglo won't make your hair look greasy and lifeless." What two words will stick in your mind when you think of Beautiglo? Or consider this sentence from a memo: "The new pension regulations are not designed to lower employer contributions at the employees' expense." Doesn't this raise suspicions in your mind? In a real situation, employees who had supported the changes might reconsider. Rewrite the following sentences to emphasize positive characteristics:

(a) WonderLawn offers all these added features, yet it doesn't cost more than franchised lawn service.

WonderLawn offers all these added features for the same price as franchised lawn service.

(b) Community college instructors are not out of touch with recent developments in their fields.

CC instructors need to be updated in their fields.

(c) Eden's Best silk plants will not wilt, turn yellow, or become diseased like real plants.

Eden's Best silk plants retain their natural beauty forever.

(d) The new parking garage will not be more expensive or less convenient than the old lot.

Convenience & cost are not an issue for the new parking garage.

(e) You won't be bored and lonely at Club Paradiso.

Club Paradiso is lively & active.

Compare your answers with those on page 46.

Changing words and phrases will help give your writing a positive rather than a negative tone. But just as the real source of a you-centred tone is genuine concern for and sympathy with the audience's point of view, so a really positive message must reflect the problem-solving approach of the writer or speaker. If you are trying to punish your audience and find excuses for not meeting their requests, or if you lack commitment to your job, it will be hard to sound positive. Good editing simply ensures that your communication reflects your positive orientation to your audience and message. This in turn reinforces a working environment that is open and committed to productivity.

Read the following memo. What do you think it says about Fred's attitude to his job?

To: Bill Smith Date: 07 06 30
From: Fred Bloom FB
Re: Faulty Desk

Would it be too much trouble for the company to get around to doing something about my desk? Three weeks ago you told me to speak to Edna about it, and I'm still waiting. I'm tired of operating with one drawer and two square feet of work space. It's hard enough to do my job, without this kind of aggravation.

Rewrite this memo to convey a more positive, cooperative tone.

_____

_____

_____

_____

_____

_____

_____

_____

Non-Discriminatory Tone

Effective communication requires respect between speaker or writer and audience. Non-discriminatory language is one way of demonstrating and fostering respect. People cannot work together productively or work effectively with customers or clients if they do not feel mutually valued as individuals, regardless of age, sex, race, disability, or any other characteristic. Thoughtless language habits can jeopardize this necessary trust. Two potential pitfalls are racism and sexism.

Racism

1. **Do not allude to someone's racial or ethnic origin unless it is relevant to the matter at hand.** The following are examples of how unnecessary racial references should be eliminated.

   Examples

   ✗ POOR
   Nelson, who came here from Jamaica eight years ago, is joining us as junior sales manager.

   ✓ IMPROVED
   Nelson is joining us as junior sales manager.

   ✗ POOR
   A Chinese woman came into the store.

   ✓ IMPROVED
   A customer came into the store.

2. **Do not give the impression that all or most people from a similar racial or ethnic background share similar characteristics, even if they are positive ones.**

   Example

   ✗ POOR
   As an Asian, Julie has adapted quickly to the wireless environment.

   ✓ IMPROVED
   Julie has adapted quickly to the wireless environment.

When planning promotions, social events, or similar activities, remember that "ethnic" themes should be very carefully thought out. You may perceive your "Mexican Night" with its sombreros and serapes, donkeys and cacti as cute and harmless, but a person from Mexico might find it a demeaning stereotype. Don't assume it's all right just because no one says anything.

Sexism

1. **Try to treat men and women equally in writing and speaking.** Do not identify women by marital and family status or by appearance unless these are relevant considerations.

   Example

    POOR
   Vivian Tam, a mother of two, is currently head of quality control.
   Tom Rossi, a systems analyst, is joining our department.

    IMPROVED
   Vivian Tam, a graduate engineer, is currently head of quality control.
   Tom Rossi, a systems analyst, is joining our department.

2. **Use the correct form of address.**

   Example

    POOR
   Tom and Mrs. Tam

    IMPROVED
   Tom and Vivian

   or

   Mr. Rossi and Mrs. Tam

   Do not assume that a married woman shares her husband's last name or uses the courtesy title "Mrs." Of course, many do, but those who don't may have strong views on the subject and find your assumption offensive. In any event, you should use a woman's full name if you have used the man's full name.

Example

✗ POOR
Bill Rivers and his wife Laura

✓ IMPROVED
Bill Rivers and his wife Laura Zawiski

or

Bill Rivers and his wife Laura Rivers

If you are unsure of the preferred form of address, use "Ms.," or omit it altogether:

Ms. Joan Solomon

or

Joan Solomon

3. **Watch for words ending in "-man."** Look for terms—such as "firefighter," in place of "fireman"—that do not make an assumption about the sex of the person holding that job.

Sexist	Non-discriminatory
workman	worker
repairman	repair person or maintenance worker
salesman	sales representative
businessman	business person
mailman	letter carrier
policeman	police officer
handyman	handy person

Titles like "cleaning lady" and "saleslady" are also sexist. The word "lady" is generally regarded as demeaning in any context, except when it is paired with "gentleman."

4. **Do not draw attention to a person's sex unless it is relevant.**

   Example

   ✗ POOR
   Jackie Giannidis, our leading woman sales rep, will be taking over in Western Canada.

   ✓ IMPROVED
   Jackie Giannidis, a leading sales rep, will be taking over in Western Canada.

4 TONE IN ORAL COMMUNICATION

In speaking, just as in writing, the words you choose will send a message about your attitude toward your subject and your audience. If you begin every sentence with the word "I" or "we," your audience will quickly come to feel that your first concern is yourself. If you speak to people in the third person—"Staff will please return to their desks"—you will create a sense of alienation rather than community. Likewise, if you begin every conversation with "I'm afraid that ... ," "I can't ... ," "We don't ... ," "You're not allowed to ... ," your listeners will soon learn to avoid you when something needs to be accomplished. The principles discussed under "You-Centred Tone" and "Positive Tone" on pages 32 and 35 are equally important in oral communication, as we will see from Gary's experience with Acme Computer.

CASE 3

Gary had recently been given a printer to hook up to his computer. It was an older model, the kind that used tracked paper, but it was clean and in working order except for the paper feeder, which had lost a few teeth on one side. Gary decided to phone the manufacturer for some help. Here is what happened:

Acme Computer (AC):	Good morning, Acme Computer.
Gary:	Hello, I have a question about a model ...
AC:	You'll have to talk to Customer Service. Just a minute ...
AC:	Customer Service, Andrea speaking.
Gary:	Hello, I have a question about a model LX-4 Jetwriter printer.

AC: I'm sorry, we don't make that model anymore. Did you say LX-4? We haven't made those since 1998.

Gary: Yes, I know, but I have one and it's in pretty good shape except for one of the sort of wheel things that pull the paper in …

AC: We call that the paper feeder.

Gary: Thank you. Yes, well one side has lost some teeth, and …

AC: You can't use the printer like that. The paper will get jammed.

Gary: Yes, I know, so I was wondering if this wheel part could be replaced.

AC: Well, we don't make that printer anymore. That's an old printer.

Gary: Well, perhaps there's another wheel that would fit.

AC: We build precision products here. You can't just put a part from one machine into another. Besides, that whole mechanism is obsolete.

Gary: Maybe the wheel could be repaired.

AC: This is a manufacturing company. We don't service computers.

Gary: Well, could you tell me who does repairs for Acme products?

AC: There are so many places, I couldn't give you all the names over the telephone.

Gary: Okay, well, I guess that's it.

AC: If you need any more help, give us a call.

When you write, you have an opportunity to revise your work. You can look for places where "I" can become "you" and where "This can't be done until …" can be changed to "This can be done when…." In speaking, you have only one chance. If you are delivering a prepared speech, of course you can edit it for you-centred, positive words. But most oral communication, even if it is planned, is not scripted word for word. Thus, your attitude comes across more obviously, without the benefit of editing. Fortunately, your writing practice will reinforce the you-centred, positive, non-discriminatory attitude you want to project.

Writing and speaking skills influence each other. In addition, you should include your goals for achieving an appropriate tone whenever you plan oral communication, whether it's a phone call or a major meeting. Try to be consciously aware of other people's choice of "I" or "you," "can" or "can't." Their example, good or bad, will also reinforce your decision to make your tone reflect a more cooperative, can-do attitude.

Tone of Voice

Tone in oral communication has the added dimension of tone of voice — the emotional quality of the sound of our speech. Tone of voice can reinforce our choice of words or totally contradict it. A warm, caring speech full of "you" and "your" will lose all credibility if it is delivered in a monotone, without eye contact. Because our earliest experiences are preverbal, we instinctively believe the messages that tone of voice and body language send us, rather than those of the words alone.

Answers to Chapter Questions

Page 36:
(a) The photocopier was not turned off.
(b) We cannot fill your order as requested.
(c) The committee did not accept the suggestion that was put forward.

Page 38:
(a) The conference room will be free at noon.
(b) Your pool is already covered by your house insurance.
(c) Your stationery order will be filled on Thursday.
(d) Pre-authorized payment will ensure that your bills are always paid on time, even when you're away.
(e) This item is available in white, grey, or tan.

Page 39:
(a) WonderLawn gives you all these added features for the same price as a franchised lawn service.
(b) Community college instructors keep up to date with recent developments in their fields.
(c) Eden's Best silk plants retain their natural beauty forever.

(d) The new parking garage will provide the convenience of the parking lot at the same price.
(e) You'll enjoy yourself and meet new friends at Club Paradiso.

CHAPTER REVIEW/SELF-TEST

1. Good business writing style means
    (a) expanding your ideas with impressively long words and phrases.
    (b) using special formal phrasing found only in business writing.
    (c) entertaining people with your literary brilliance.
    (d) all of the above.
    (e) none of the above.

2. A message is clear if your intended audience
    (a) agrees with everything you say.
    (b) receives your ideas quickly and accurately.
    (c) is impressed by your expertise and large vocabulary.
    (d) enjoys reading or listening to it.

3. Make your message clear to your intended audience by
    (a) repeating each point for emphasis.
    (b) writing exactly as you speak.
    (c) being concise and using simple, concrete words.
    (d) searching for new words in a thesaurus.

4. Grammar errors can make your message unclear by
    (a) distracting the reader or listener.
    (b) allowing it to be interpreted two different ways.
    (c) creating a conflict between common sense and what the message actually says.
    (d) all of the above.

5. The tone of a message refers to
    (a) its emotional impact on the audience.
    (b) the educational level of the speaker or writer.
    (c) whether it is formal and correct.
    (d) whether it is good news or bad news.

6. Under which circumstances would you avoid using the word "you"?
    (a) You want your audience to follow instructions or procedures.
    (b) You want your audience to approve an idea or decision.

(c) Your message draws attention to a problem or mistake.
(d) You are selling a product.

7. Which of the following is **not** a reason why negative messages are less effective than positive messages?
    (a) Negative messages are less clear.
    (b) Negative messages can make the audience feel resentful and uncooperative.
    (c) Negative messages turn people off.
    (d) Negative messages raise people's suspicions.

8. Language is non-discriminatory when it
    (a) says nice things about everybody.
    (b) treats ladies, the elderly, and the disabled with special respect.
    (c) points out positive characteristics of minorities.
    (d) treats all equally without reference to membership in any group.

9. It is appropriate to mention someone's ethnic origin in the workplace if
    (a) you are curious about what it is.
    (b) you have had a positive experience with other people who share that background.
    (c) it is clearly relevant to a work-related issue.
    (d) you enjoy food from that country.

10. It is appropriate to mention someone's family status in the workplace if
    (a) he or she is heterosexual.
    (b) he or she is middle-aged.
    (c) he or she is wearing a ring.
    (d) it is clearly related to a work-related issue.

Answers

1.E 2.B 3.C 4.D 5.A 6.C 7.A 8.D 9.C 10.D

EXERCISES

1. Rewrite the following sentences to make them clear and concise:
    (a) This paper is the kind of paper that looks more expensive than ordinary paper, yet doesn't cost more than ordinary paper costs.
    (b) While the construction crew is refinishing the floorboards they will be inspected for termites.

(c) At this point in the year individuals are often experiencing difficulty in the cash flow area with reference to back-to-school expenses.
(d) After removing all the clutter the office looked more professional and efficient.
(e) Suits that are double-breasted do not look good on men who are short.
(f) There are several aspects of the salary situation which suggest an upward trend.
(g) A refund in the amount of $58.95 has been credited to your account, as per your request.
(h) It has come to my attention that the area of parking fees has experienced a falling-off in revenue.
(i) I would be prepared to buy a house from an independent contractor, if well-built.
(j) It is expensive to rent an apartment in the city of Vancouver.

2. Rewrite the following sentences to make them positive, reader-oriented, and non-discriminatory:
   (a) Obviously, you failed to calculate the bank deposits accurately.
   (b) These are not the cheap, mass-produced hairpieces that always look fake.
   (c) We have a wide assortment of styles in stock.
   (d) Workmen will be installing the carpet on Tuesday.
   (e) Customers will please refrain from leaving valuables in their lockers.
   (f) Get one of the ladies in the cash office to handle it.
   (g) You won't be disappointed by the food and service at Antonio's.
   (h) Our district manager, a girl with years of experience, handled your account personally.
   (i) We were delighted to have the opportunity to answer your questions about Gemstone wall coverings.
   (j) Emerald Weave outdoor carpeting won't rot or fade like other carpets.

3. You work for the head office of Dominion Suites, a hotel chain. Two weeks ago your supervisor asked you to prepare a report on "Dominion Plus Points," the reward program offered to guests who stay frequently at Dominion Suites hotels. Your job was to compare the benefits your program offers to its members with the benefits offered by your top three competitors. Here is the first draft of your opening paragraph:

   I am happy to report that Dominion Plus Points is still ahead of the competition. However, I have to point out that our rivals are doing everything they can to close the gap. The study I recently completed shows that many of the benefits we introduced last year have been imitated by other reward programs. Let me stress that most of our rewards have been items like free accommodation and room upgrades that appeal to business travellers. Although this is an important target market, I want to emphasize that more and more customers have expressed an interest in consumer items like sports equipment and electronics. Because of this trend, I want to present some recommendations that Dominion Suites might pursue.

Rewrite this paragraph to make it more "you-centred."

4. Read the following dialogue. Identify statements that would be likely to alienate the listener. Be prepared to improve them in a role-play exercise.

Deborah has been walking around the "separates" section of a large department store for several minutes. Finally she approaches Lisa, a salesclerk.

Deborah: Excuse me, I wonder if you could help me …

Lisa: Are you looking for something?

Deborah: Yes, I need a blouse to go with a grey tweed skirt—perhaps in a pale blue, or maybe yellow.

Lisa: Well, I don't think we have any blue blouses. If we do, they're over there.

Deborah: Perhaps you could show me.

Lisa: Well, there's what we have. What size are you looking for?

Deborah: Ten.

Lisa: Most of the tens are gone. Here's something. *(She takes it off the rack.)*

Deborah: Okay, I'll try that. And maybe this peach one.

Deborah: *(returns from change room)* I think the blue one fits a bit loosely.

Lisa: Well, that style of blouse doesn't really suit you; that collar doesn't look good if you don't have much up top. What about the peach?

Deborah: It's a bit too orange for me …

Lisa: Yeah, if your skin has a greenish tint, that shade really brings it out.

Deborah: Does it come in another colour?

Lisa: Our colour selection is pretty limited. There's this raspberry shade …

Deborah: That's nice.

Lisa: Yeah, a lot of older women like this because it sort of brightens up your face.

Deborah: I see. Would it go with grey tweed?

Lisa: Yeah, it wouldn't look too bad. Depends on your taste.

Deborah: Well, I think I'd like to take it home and try it on with the skirt. Can I bring it back if I don't like it?

Lisa:	There are no exchanges unless you bring it back within ten days with your receipt.
Deborah:	That shouldn't be a problem. If you ring it up, I'll write a cheque.
Lisa:	We only accept cheques with a printed address and two pieces of ID.
Deborah:	That's not a problem. *(She starts to write.)* What's the total?
Lisa:	It's on the bill. $41.68.
Deborah:	Well, thank you. Bye.
Lisa:	Bye. I hope it doesn't clash with your skirt.

PART TWO

COMMUNICATION STYLES

CHAPTER THREE

PRESENTING WRITTEN MESSAGES

Chapter Objectives

Successful completion of this chapter will enable you to:

1. Produce an attractive, well-spaced letter using a standard letter format.

2. Present a memorandum using a standard format.

3. Prepare a message for e-mail transmission.

4. Lay out a written message for transmission by facsimile machine.

INTRODUCTION

First impressions are important in the working world. The layout and general appearance of a written message start to communicate even before the first word is read. This chapter shows you how to present letters, memos, and electronically transmitted written messages so that they make a positive first impression.

1 LETTER PRESENTATION

Before you begin to organize the content of a letter, you need to think about its layout and general appearance. With many options available for sending written messages, a letter sent through the postal system is still the usual choice for an important, formal message. A letter is also the best means for establishing written communication with a new or potential customer or business associate, before fax numbers or e-mail addresses are exchanged. Many people will not open e-mail messages from unknown senders. Thus it is particularly important that every element of your letter look its best.

Paper

1. Invest in top-quality white bond paper and matching envelopes. Ask a stationery store to help you find the right weight or thickness of paper, keeping in mind the requirements of your printer.
2. Never use letterhead paper or envelopes from your workplace for personal correspondence. This practice sends a very negative message about the writer's regard for other people's property. Use letterhead only when you are writing on behalf of that organization.
3. If you are starting your own business, get professional help in designing your letterhead. The expense per sheet will be low and, in any case, well worth it for the boost an expert design gives to your company's credibility.
4. Use letterhead paper only for the first page of a letter. Type or print the following pages on plain, matching bond.

Spacing

1. Plan your message to ensure that it is well placed on the page. Ideally, the left and right margins should be of equal width, and so should the top and bottom margins. No matter how short your letter, it should come below the midpoint of the page to look balanced.
2. Do not carry over to a second page unless you have at least two more lines of text before the closing.

3. A business letter should be divided into short paragraphs with lots of white space in between. This invites the reader into the page, instead of hitting him or her with a wall of text.

Typeface (Font)

1. Use a standard typeface such as Courier or Helvetica, 10 or 12 points. Aim for a conservative "typed" appearance. Save script and other fancy fonts for rock band flyers.
2. Never mix fonts in a letter. Do not "cut and paste" unless the font and size are consistent.

Elements of a Business Letter

A business letter consists of four parts: heading, opening, body, and closing. Certain rules, or, more accurately, conventions, dictate how each part of a business letter should be presented. These conventions change, just as conventions in business attire change—slowly. Careful attention to small details will make the difference between an attractive, functional, and professional-looking letter and one that reflects the writer's lack of knowledge and experience.

Heading

Letterhead stationery requires only the addition of the dateline to complete the heading. Type the date at least two lines below the letterhead and four lines above the receiver's address. Use either the traditional form:

February 28, 2007

or the newer all-numeric date:

2007 02 28

If you are using plain paper, put your return address at least 3 cm below the top of the page. **Do not put your name in the heading.** Type the date immediately under the return address.

Opening

The **receiver's name and address** direct the letter to its intended reader and provide you with a reference on your file copy. Include the receiver's courtesy title (Miss, Mrs., Ms., Mr., Dr., Prof.), if known; the receiver's name; job title, if known; organization name; and full mailing address. See Figures 3.1, 3.2, and 3.3 for the position of each item.

The **salutation** greets the reader personally. If your reader's name is, for example, Ms. Leslie Khan, you may begin "Dear Ms. Khan," or, if you are on friendly terms, "Dear Leslie."

If you are not using a courtesy title such as "Mr." or "Ms." in the receiver's address, omit it in the salutation: "Dear Leslie Khan," or "Dear Leslie," if you know your reader well. The practice of omitting courtesy titles is becoming more popular. It is particularly helpful if you do not know which of Miss, Mrs., or Ms. would be most acceptable, or whether Chris Suarez is a man or a woman.

The proper form of salutation is "Dear Courtesy Title Family Name," or if you are not using a courtesy title, "Dear Given Name Family Name." **Never use a courtesy title with both names.**

Example

✗ POOR
Ms. Rose Chu
38 Wynford Dr., Apt. 1302
North York ON M5P 1J3

Dear Ms. Rose Chu

✓ IMPROVED
Ms. Rose Chu
38 Wynford Dr., Apt. 1302
North York ON M5P 1J3

Dear Ms. Chu

or

Rose Chu
38 Wynford Dr., Apt. 1302
North York ON M5P 1J3

Dear Rose Chu

If you do not know your intended reader's name it is better to omit the salutation. "Dear Sir or Madam" sounds very old-fashioned. If the letter is short you could begin immediately with the first sentence of the body of the letter, as in Figure 3.2. Or, you could use an **attention line,** directing the message to a particular department or to someone with a particular job title.

Example
Best Buy
20 City Centre Drive
Brampton ON N5G 1E9

Attention: Customer Service Manager

Another alternative to a salutation is a **subject line.** The subject line summarizes the contents of your letter and helps a receptionist or mailroom employee direct it to the appropriate reader.

Example
Great Lakes Insurance Company
34 Confederation Way
Thunder Bay ON N4G 1H6

Subject: Liability Insurance Rates

Example
Box 423, Report on Business
444 Front St. W.
Toronto ON M5V 2S9

Subject: Job File 93-701

Do not use "To Whom It May Concern." How can your readers know if the message concerns them until they have read it? This phrase simply wastes everyone's time.

Body

A letter with a salutation may also have a subject line. This can be a useful way of focusing the attention of a busy person who receives and files a lot of correspondence.

Example

Dear Mr. Ackerman
Subject: Applications for Youth Employment Grants

Subject lines are not necessary for letters to private individuals.

Lay out the body of the message in short paragraphs. Follow the spacing guidelines on page 61.

Closing

If you have used a salutation, you should end your letter with a **complimentary closing.** Formal complimentary closings include "Yours truly," "Sincerely," and "Yours sincerely." Note that only the first word is capitalized. Informal closings include "Good luck," "Best wishes," and so on. Omit the complimentary closing if you began your letter with a subject or attention line instead of a salutation. After the complimentary closing, or immediately after the body of the letter, if you are omitting the salutation and closing, comes the **signature block.**

Leave at least three blank lines for your handwritten signature (more if your writing is large). Type your name, and include your job title beneath your name if you are writing on behalf of your organization. Normally, courtesy titles are omitted in the typed name. You may prefer to use your courtesy title if your name is one that could be either a man's or a woman's (Kim, Terry, Pat). A woman who has strong feelings about being addressed as "Miss" or "Mrs." may wish to include the title she prefers in her typed name. If there is no title, it is generally assumed that she prefers "Ms." Always omit your courtesy title in your handwritten signature.

Examples

Sincerely

Ravi Edwards-Singh

Ravi Edwards-Singh
Manager, Accounts Receivable

your initials

Sincerely yours

Kim Greenough

Mr. Kim Greenough
President

Best wishes

Maureen Laird

Maureen Laird, Director
Food Services

ESL TIP

In English-speaking countries, people always introduce themselves and sign their names in the order Given Name Family Name. For example, three members of the Burgess family might introduce themselves as Tanya Burgess, James Burgess, and Matthew Burgess. This may not be the order you are accustomed to, but it is important to follow this convention when communicating in English. Otherwise, people will unintentionally address you by your family name when they are intending to establish a friendly "first name" relationship. They will be unable to find your name in a list alphabetized by family name, such as a telephone directory. Messages and files will be misplaced.

The form Family Name, Given Name or FAMILY NAME Given Name (e.g., Burgess, Tanya or BURGESS Tanya) might appear in a list or extract from a list. The comma or spacing tells the reader that normal English-language name order is not being used. This form is not appropriate for correspondence.

Letter Format

The format of a letter refers to the arrangement of the four elements on the page. The two letter formats most commonly used in business today are the Full Block and the Modified Block. In the **Full Block** format (Figure 3.1), every line begins at the left margin. This format is becoming the preferred choice in most businesses

FIGURE 3.1 Full Block Format: Plain Stationery

> 95 Glencairn Drive
> Hamilton ON L4B 6M4
> 2006 11 23
>
> Mr. Rene Laplace, President
> Mohawk College of Applied
> Arts and Technology
> 2951 Fennell Ave. W.
> Hamilton ON L9C 5R2
>
> Dear President Laplace
>
> Thank you for participating in our class symposium on Management in the Public Sector. Your varied examples from your own wide range of experience were very helpful in illustrating the points we were hoping to explore.
>
> With your permission we would like to post the text of your opening remarks on our website. In the meantime you might enjoy seeing some pictures from the event at www.collegebusnet.ca.
>
> Thank you again for helping to make our symposium a highlight of the semester.
>
> Sincerely
>
> *[signature: Michael Swoboda]*
>
> Michael Swoboda, Convener
> Symposium Organizing Committee

because it is the easiest to keyboard. You do not have to worry about lining up the return address with the signature block or indenting the first line of each paragraph. The Full Block format generally has **open punctuation,** which means that only the body of the letter is punctuated. There is no punctuation after the salutation or complimentary close. On a letterhead, the formatting rules are the same. The only difference, as Figure 3.3 shows, is that you need not type the return address.

The **Modified Block** (Figure 3.2) is somewhat more old-fashioned. The return address, date, complimentary close, and signature block are placed at the halfway point across the page. The first line of each paragraph is not indented. A letter in Modified Block format may use open punctuation or **mixed punctuation.** Mixed punctuation means that the salutation is followed by a colon ("Dear Mr. Healey:") and the complimentary close by a comma ("Sincerely,"). Unless your organization has a standard format, you can choose either the Full Block or Modified Block, open or mixed punctuation; just follow your preferred style consistently. Note the presentation of a modified block letter in Figure 3.4.

FIGURE 3.2   Modified Block Format: Plain Stationery

```
                                              1281 Grand Blvd.
                                              Oakville ON  L6N 2F6
                                              2007 06 19

Optima Corporation
28 Taylor Road
Toronto ON  M2A 3T4


Please send me a copy of your latest catalogue. I am enclosing a cheque for
$4.00 and a stamped, self-addressed envelope.


                                              Winsome Lincoln
                                              Winsome Lincoln
```

FIGURE 3.3 Full Block Format: Letterhead Stationery

Insulate Inc.
22 Sussex Drive Ottawa Canada   K1S 1A5

2006 01 06

Mr. Albert Fleury
82 Waterloo Ave.
St. John's NL A1P 6Y2

Dear Mr. Fleury

Here are the brochures you requested on home insulation. On the back cover of "Conserving Home Heat," you will find a list of installers who can provide you with information about specific products. Good luck with your renovation plans.

Sincerely

[signature]

Susan Giordano
Consumer Information Officer

Use the lines provided on the next page to show the placement in the Full Block format of the following elements:

heading {
1. return address (use your home address)
2. date (today's date)
}
opening 3. receiver's name and address (president of your college) + greeting
4. complimentary close
5. signature block

Compare your letter with the one in Figure 3.1.

Correct the errors in the following according to the guidelines given in this chapter. Circle the errors and write the correct version beside each example.

Heading

1. Michael Bowden, Manager
   Aug. 23rd, 2008

   1. — Should be address — not a name
      August 23, 2008

2. 2006 15 01
   Teresa Perez
   16 Wellesley Crt.
   Guelph, ON N2R 8A9

   2. 2006 01 15
      — name not in heading
      — address before date

Opening

3. Western Metallurgical Industries
   Calgary, Alberta
   ~~Dear Sir:~~ never use, put a subject line
   Subject: Request for Survey Results

   3. not a full address

4. Sales Manager, Monte Carlo Sportswear
   218 Spadina Ave.
   Toronto, ON M6G 2W4

   Attention: Ms. Barbara Dimopoulos
   Dear ~~Barbara~~ Ms Dimopoulos

   4. _____

Closing

5. Sincerely:
   Mr. Chris Yeung
   Mr. Chris Yeung, Assistant Account Director

   5. Sincerely,

6. Yours ~~Sincerely~~ sincerely,
   Betty ~~~

   6. _____

FIGURE 3.4 Modified Block Format: Letterhead Stationery

SUPREME WIRE & CABLE
298 B.C. Place Drive   Vancouver BC   V2A 3B3   (604) 218-5252

November 15, 2007

Mr. Norman Cheung
Senior Sales Representative
Niagara Industrial Cleansers
234 Lake St.
St. Catharines ON   L6R 4S3

Dear Mr. Cheung:

After testing the solvent samples you sent us last week, we have some questions:

1. Is there any difference in flammability among the three samples?

2. Are all the solvents compatible with your K-900 grease foam?

3. Can you supply the Cascade-30 solvent in 60-litre drums?

Your early reply will enable us to complete our December order.

Sincerely,

Neil Hall, Director
Purchasing

FIGURE 3.5 Addressing an Envelope

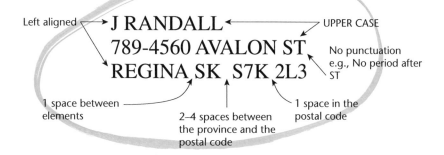

TABLE 3.1 Canadian Provinces and Territories

English Name	Symbol	French Name
Alberta	AB	Alberta
British Columbia	BC	Colombie-Britannique
Manitoba	MB	Manitoba
New Brunswick	NB	Nouveau-Brunswick
Newfoundland and Labrador	NL	Terre-Neuve-et-Labrador
Northwest Territories	NT	Territoires du Nord-Ouest
Nova Scotia	NS	Nouvelle-Écosse
Nunavut	NU	Nunavut
Ontario	ON	Ontario
Prince Edward Island	PE	Île-du-Prince-Édouard
Quebec	QC	Québec
Saskatchewan	SK	Saskatchewan
Yukon	YT	Yukon

Source (Figure 3.5 and Table 3.1): © Canada Post Corporation, 2005. Reproduced with the permission of Canada Post Corporation. Parties interested in obtaining current information for mailing purposes should consult Canada Post Corporation's website www.canadapost.ca.

2 MEMORANDUM PRESENTATION

When a written message is sent within an organization it is called a memorandum, memo for short, rather than a letter. Most organizations use standard memo forms or computer macros. These may be mass-produced for use by any company, like the following memo,

---

INTER-OFFICE MEMO

TO: DATE:
FROM:
SUBJECT:

---

or they may be specially produced for one organization. In any case, a standard form ensures that anyone reading or filing a memo in that organization will know where to look for information about the contents of the memo, who sent it, and for whom it was intended. This saves reading, filing, and access time. Regardless of the order in which the headings are placed, each line has an important function.

To:

This line corresponds to the envelope address, the inside reader's address, and the salutation of a letter. Routine paper memos are not put in an envelope, other than perhaps a reusable mailer. The **To** line should thus have all the necessary identifying information, for example:

To: All Employees
To: All Warehouse Staff
To: All Part-time Cashiers
To: Tina Bandi, Director, Staff Development

[handwritten annotation: include title if known + full name]

Using your reader's title makes your memo more formal. This is appropriate if you are writing to someone you haven't met, or rarely meet, or someone senior to you in the company. Always use your reader's title if the memo is going to file, in case the original reader leaves or changes jobs. Always use your reader's full name. These practices prevent mix-ups and are valuable for future reference.

From:

This line takes the place of the complimentary close and signature block. Some writers like to sign their memos, but expressions like "Sincerely" are always omitted. The preferred method of giving personal authorization to a paper memo is to write your initials beside your typed name, like this:

TR

From: Toby Rich, Supervisor—Accounts Receivable

Subject: *or* **Re:**

This line is an important part of your message and requires some thought. The subject line motivates your readers to read the memo, focuses their attention for better reading comprehension, and helps them find the message for later reference. In order to do this the subject line must be specific, clearly identifying the contents of the memo.

POOR
Subject: Meeting

IMPROVED
Subject: Budget Meeting May 8

Date:

You may use the traditional month/day/year form:

October 7, 2007

or the newer all-numeric form:

2007 10 07

use your intitials at bottom

3 E-MAIL PRESENTATION   *less secure*

E-mail offers you the speed of the telephone and the permanence and convenient access of a written message. You can use it for internal correspondence, in place of a paper memo, and to communicate with outside organizations, customers, or clients. At the moment, e-mails are regarded as less formal than letters. E-mail messages are certainly less secure and should never be used to transmit information you wish to keep confidential. Employers and fellow employees may be able to legally access messages, even deleted ones, sent to or from your workplace computer.

treat email like a mini letter — greeting + ending

When you send or reply directly to an e-mail you do not have much control over the appearance of your message, as it may change in transmission. If the layout of the contents is important, or if you expect that the message will be printed and circulated or kept for reference, you should send it as an attachment, assuming your receiver knows you and will be willing to open an attachment. An attachment will give you all the advantages your computer offers in creating a well-laid-out page. Letter format is not necessary, but be sure to include your name on the attachment so that it can stand alone without the transmitting e-mail.

If you are sending a brief message or carrying on an e-mail dialogue, an attachment is not necessary. It is always a good idea to use your receiver's name, if known, at the beginning of the message and to include your own at the end, even though this information appears on the frame. It adds a "you-centred" touch and is helpful if your receiver wishes to forward the message.

If you will be communicating with organizations from your home server—for example, applying for jobs or obtaining information—it is best to avoid Yahoo! and Hotmail addresses and screen names like lovrrgrrl and L'il_Buffy as many spam filters will screen out or quarantine your messages.

FIGURE 3.6 Sample E-Mail Screen Layout

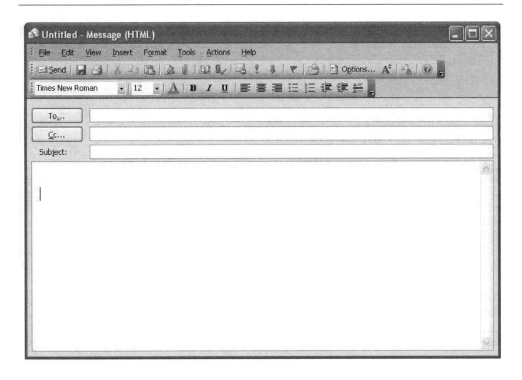

The subject line is the most important part of an e-mail. It should help your reader to open your message promptly, file or forward it appropriately, and find it again when it is needed. Make it clear and specific. Be particularly careful to avoid any phrases that could be mistaken for "spam"; for example, "Important New Offer," "Management Action," or "Latest Positions."

4 FACSIMILE MESSAGE PRESENTATION

The most efficient use of a fax machine is to send messages that already exist in print; for example, you might fax your accountant a copy of a letter you received from the Canada Revenue Agency, or fax a floor plan of your office to a supplier of office furniture. If you have to prepare a message from scratch for immediate transmission, it makes more sense to send it directly from your computer by e-mail. You may have to prepare a message specifically for facsimile transmission if you wish it to arrive as soon as possible and you are in one of the following situations:

(a) You or your intended receiver do not have access to a computer with e-mail capacity.
(b) You have the receiver's fax number, but no e-mail address.
(c) You are faxing other documents and need to attach a further message to the receiver.
(d) You have been told to communicate by fax.
(e) You have been told to communicate by letter, but must use a faster method to meet a deadline.

In these situations, or any time you expect your receiver to reply by letter mail, you can put your message in letter format and then fax it. If you expect a faxed reply, an e-mail reply, or no reply, you can simplify your layout by omitting information that is only relevant to the postal system. Most companies provide a fax cover sheet (see Figure 3.7), which can contain an informal message, or direct attachments or a more formal message to the right person. Brief answers to a faxed message can be handwritten on the bottom of the original message, if this is convenient. A handwritten message shows that you gave an urgent message your immediate personal attention.

When a fax is sent, the sender's fax number is automatically printed on the receiver's copy. A business fax usually prints the company name as well. Using your employer's fax machine for personal correspondence (like a job application) sends the same message as using your employer's stationery: "I am a thief."

FIGURE 3.7 Sample Facsimile Cover Sheet

Seneca College
FAX Transmission Cover

Transmit To: **A. E. NEWMAN, PRES.**
Company: **ACCC**
FAX Number: **(416) 555-1212**
Pages (Incl. Cover): **6**

Message: *I am sending you a revised copy of my report. The changes you requested have been made. Give me a call if you have any further changes; otherwise I will send it to the committee tomorrow.*

Originator: **L. BLOTZ**
Department: **PRESIDENT'S AREA**
Telephone (416) 491-5050 Extension **2000**

CHAPTER REVIEW/SELF-TEST

1. Under which circumstance would a letter **not** be a good choice for sending a written message?
   (a) The message is important and formal.
   <u>(b)</u> The message needs an immediate response.
   (c) You are communicating with someone for the first time.
   (d) All of the above.

2. A business letter has the following parts, in the following order:
   (a) opening, body, closing.
   (b) heading, body, closing.
   <u>(c)</u> heading, opening, body, closing.
   (d) opening, heading, body, closing.

3. Space your letter to ensure that
   (a) it is all on one page.
   <u>(b)</u> some part of it comes below the midpoint of the page.
   (c) the right margin is wider than the left margin.
   (d) you have at least two lines on which to write your signature.

4. Always use a salutation in a letter unless
   (a) you do not know the receiver's name.
   (b) you do not expect a reply.
   (c) the letter is friendly and informal.
   (d) you are writing on behalf of an organization.

5. Do not use a complimentary closing unless
   (a) you are on friendly terms with the receiver.
   (b) you have lots of room at the bottom of the page.
   (c) you are writing on behalf of an organization.
   (d) you opened with a salutation.

6. Sending a message within an organization in memo format ensures
   (a) that important information identifying the sender, receiver, and subject is included.
   (b) that the message looks authorized and work-related.
   (c) that important information about the sender, receiver, and subject is easy to find.
   (d) all of the above.

7. Under which circumstance would an e-mail **not** be a good choice for a written message?
   (a) It is important for the message to look clear and professional.
   (b) You have never met the receiver personally.
   (c) The contents of the message are confidential.
   (d) The message does not need an immediate response.

8. Use an attachment to control the appearance of your e-mail message on the page unless
   (a) your receiver might be unable or reluctant to open attachments.
   (b) your message is brief and informal.
   (c) you and your receiver are sending related messages back and forth.
   (d) all of the above.

9. The best way of avoiding spelling and grammar errors, which could spoil the impact of your e-mail messages, is by
   (a) proofreading on the screen before you send the message.
   (b) printing the message and proofreading the hard copy.
   (c) using the spelling and grammar tools on your computer.

10. The fax machine is a good choice if you wish to
    (a) forward a document that already exists in hard copy.
    (b) create a new message for immediate delivery.

(c) send a confidential message.
(d) ensure the professional appearance of your message.

Answers
1. B 2. C 3. B 4. A 5. D 6. D 7. C 8. D 9. B 10. A

EXERCISES

1. Number the items below in the order in which they appear in a business letter. In the space provided, write an example of each item.

   __5__ complimentary closing

   __4__ salutation

   __3__ receiver's address

   __1__ sender's address

   __6__ signature block

   __2__ date

2. Proofread the following letter, correcting all errors in presentation. Rewrite on plain stationery.

> ~~Lisa Crawford~~
> 418 Lynford Cres.
> Toronto, ON M4Q 2P3
>
> Oct. 28, 2008
>
> Mrs. Doreen Park
> Park Floral Designs
> [street address]
> Windsor, ON [postal code]
>
> Dear Mrs. ~~Doreen~~ Park:
>
> Thank you for the estimates you provided for our wedding flowers. Unfortunately, the engagement has been called off, and we will not be requiring your services. I will keep your brochure ~~on hand~~ for future reference.
>
> Sincerely,
>
> *Lisa Crawford*
>
> Student

3. (a) Reformat the following message for facsimile transmission.

To: Doreen Park
Park Floral Designs

Subject: Wedding Flower Estimate

Home Care Associates
4140 Main St. Ste. 307
Hamilton ON L2G 1H9
May 14, 2009

Mr. Lawrence Washburn
1289 Broadway Ave.
Hamilton ON L2G 1H9

Dear Mr. Washburn

Thank you for your interest in Home Care. Here are the answers to your questions:

1. The services of Home Care personnel are fully paid for by the Ministry of Health if they are considered medically necessary physical care. This recommendation can be made by any of the professionals listed on the attached application.

2. Other services are provided on a user-pay basis. Financial assistance may, however, be available from other agencies such as Veterans' Affairs or Regional Social Services.

3. An appointment can be made to assess your circumstances as soon as an intake worker has opened a file for you. To do this, we need your Health Card and SIN numbers.

We look forward to hearing from you soon. Home Care can usually be put in place within two weeks of your intake appointment.

Sincerely

Tracy Boateng

Tracy Boateng
Administrative Assistant

(b) Assume that you are the receiver of this message. Write a reply at the bottom that could be faxed back to the sender.

4. Proofread the following memo for errors in presentation. Create a suitable subject line.

> TO: Imram   DATE: Tues.
>
> FROM: Al
>
> SUBJECT:
>
> We need to get together to prepare an agenda for the March 4 meeting. If you could leave a copy of your timetable in my mailbox, I'll get back to you with some possible times.
>
> Sincerely,
>
> Al

5. Identify problems in presentation that might prevent the following messages from achieving their purpose:

(a)

> **SUBJECT:** RE: Problems with my mark
> **FROM:** Katy Rogers katy.rogers@collegenet.on.ca
> **DATE:** Fri, 07 May 2006 11:38:25 -500
> **TO:** Professor Larry Enfield lcenfield@collegenet.on.ca
>
> hi Larry,
>
> I thought i did well good on the exam. Why did I get a F in the course?

(b)

SUBJECT: What was done in meeting today!!!
FROM: "shanellegreene greene" shanellegreene@hotmail.com
DATE: Wed, 12 Mar 2006 19:53:36 +0000
TO: anna.tang@scotiabank.com
Hi,
I was wondering if you can just
reply
tom me what was done in the staff meeting
today
because I could not make it due to not feeling
well :(...One more thing
when
are we supposed to file our RRSP journals ...I
will
appreciate it is you can send this message back
as
soon as possible.

(c)

SUBJECT: Let's get together
FROM: Diane Aloe dcaloe@admin.cora
DATE: Tue, 09 Sep 2008 14:32 -500
TO: Floor Managers floormanagers@admin.cora

IT'S TIME TO GET TOGETHER
A meeting to evaluate the summer sale period July 15-August 15 will take place Friday afternoon in the Board Room.
MEETING SEPTEMBER 12, 2:30 PM BOARD ROOM 1408
Please call me at x3129 if you have any questions or concerns.

SEE YOU THERE!

CHAPTER FOUR

ROUTINE MESSAGES

Chapter Objectives

Successful completion of this chapter will enable you to:

1. Write a message requesting information.

2. Write a message placing an order.

3. Write a positive reply to an order or a routine request.

4. Request information, place an order, and take routine requests and orders in person or over the telephone.

INTRODUCTION

Most of the communicating you do on the job will involve asking for or giving information. Because the flow of information is vital to a successful organization, it is important to master the simple techniques that will help you handle inquiries and replies quickly and clearly. In this chapter, we concentrate on routine requests and affirmative responses. Special requests are dealt with in Chapter 7, "Persuasive Messages," and negative replies are discussed in Chapter 6, "Bad News: The Indirect Approach."

The contents of a routine request or reply follow the **direct order.** This means that the **main idea** is placed first, **supporting details** or necessary information follow the main idea, and a brief **goodwill close** ends the message.

1 MAKING A ROUTINE REQUEST

A request is most likely to be answered if the reader can easily discover what information you want. If your request is vague, incomplete, or confusing, your audience will have a difficult job. You may face delays while you clarify your request. Here are some suggestions for writing clear inquiries that get the answers.

1. **Identify the purpose of your message in the first line.** Be as specific as possible. This will ensure that your message is read by the appropriate person and that it will focus his or her attention on your request. Don't let your reader get lost in lengthy introductory formulas.

   Examples

   ✗ POOR
   I was looking at some old reports the other day and I happened to notice that you served on the United Way Committee at one point and it occurred to me that you might have some information I have been trying to locate.

   ✓ IMPROVED
   Do you have the name of the United Way coordinator in York Region?

   ✗ POOR
   I am currently shopping around for a monitor and I noticed in a back issue of a computer magazine that you have some equipment I might be interested in.

   ✓ IMPROVED
   I would like some technical specifications for your XC LCD monitor advertised in the November issue of *Hardware Central*.

2. **Add any details that will help your reader help you.** The more specific your request, the more likely you are to get the information you require.

Example

POOR
I would like some information about Quebec City.

This request puts no limit on the potential contents of the reply. The sender could easily receive an armload of brochures by return mail, and perhaps none of them would contain answers to the writer's real questions.

IMPROVED
Please send me some information on hotel packages during the Quebec Winter Carnival.

This focuses the request and saves time for the sender and receiver. Do not include details that will not enable your audience to give a better response—for example, your motivation for writing, or background details about yourself or your company. Test every statement by asking yourself, "Will this help my reader answer my question?"

3. **Use questions rather than statements.** Write "Is your restaurant accessible for people in wheelchairs?" rather than "I was wondering whether your restaurant is accessible for people in wheelchairs." The question mark acts as a prompt to your reader that a reply is required. Questions also help your reader to check back to make sure that he or she has told you everything you want to know. Number your questions if you have asked more than two.

Example

THIS MUD'S FOR YOU COFFEE SERVICE:
After reading your brochure, we still have some questions about your office coffee service:
1. Does your "per user" rate distinguish between full- and part-time employees?
2. Does your "per kit" rate require the purchase of a minimum number of kits per month?
3. Can items such as cups and filters be ordered separately, or only as part of a kit?

4. Use lists or a table to display complex information.

Example

✗ POOR

Recently we received an invoice from you, number G84669, for $120.00 for two and a half hours of photocopy repair service on our Clearox L-600 at $48.00 per hour. In addition, we later needed four hours of repair service, presumably at the same rate. This was invoiced on invoice number G90421. The first invoice was dated October 7, and the second November 2. But the second invoice was for $212.00. Since the rate presumably remained the same, we are enclosing a cheque for $312.00 as payment in full.

✓ IMPROVED

We have received two invoices for photocopier repair totalling $332.00. We believe the amount should be corrected to read $312.00. Here is the calculation we used.

Invoice Number and Date	Hours	Rate	Total	Amount of Invoice
G84669 October 7	2.5	$48/hr	$120.00	$120.00
G90421 November 2	4	$48/hr	$192.00	$212.00
			$312.00	$332.00

Some e-mail programs do not transmit tables and columns, even though they appear on your screen as you are creating the message. It is safer to use a series of parallel sentences separated by blank lines.

Example

Inv. # G84669   Date 10 07   Hrs. 2.5   Rate $48   Total $120   Inv. $120
Inv. # G90421   Date 11 02   Hrs. 4   Rate $48   Total $192   Inv. $212

5. Use an action close.
An action close motivates your reader to reply promptly by setting a target date.

Examples

We would appreciate having this information in time for our December 16 meeting.

If a date is not appropriate, offer some other form of motivation.

I am looking forward to visiting my nearest Burger Barn as soon as I hear from you.

Please let me know as soon as possible so that I can complete my travel plans.

It is not necessary to thank your reader if supplying information is part of his or her routine job. If you are asking for some extra effort—for example, answers to many detailed questions, or information in situations where there is no apparent benefit to your reader (if, for instance, you are not a fellow worker or a potential customer)—you may want to add a sentence such as "Thank you for your help" or "I appreciate your assistance." Do not thank your reader for taking the time to read your letter; this is false humility of the "pardon me for living" variety, and it will make your message sound very unprofessional.

Read the advertisement in Figure 4.1. Using the Modified Block format, use the lines following it to write a routine request for at least three items of information about Sea Breeze Vacation Homes. Compare your letter with Figure 4.2.

FIGURE 4.1   Advertisement—Sea Breeze Vacation Homes

YOUR HOME AWAY FROM HOME ... IN SUNNY ORLANDO

Come to Sea Breeze and enjoy live-in convenience only 10 kilometres from Orlando's popular theme parks. Choose a three, four, or five bedroom home in a residential neighbourhood, all with

- Private outdoor pool
- Secure parking
- Satellite TV
- Internet access
- Full kitchen
- Multiple bathrooms
- Laundry facilities

★ Prices as low as $65 plus tax per night for stays of five nights or more. ★

FOR INFORMATION, WRITE

Sea Breeze Vacation Homes

235 Vista Blvd. N.
Orlando FL   34746
Or visit www.seabreezeorlando.com

FIGURE 4.2 Letter to Sea Breeze Vacation Homes

1233 63rd Avenue
Edmonton AB T6J 2E5
Canada
May 5, 2007

Sea Breeze Vacation Homes
235 Vista Blvd. N.
Orlando FL 34746
USA

I would like more information about your vacation homes advertised in the May issue of *TravelValue*. Specifically, please answer the following questions:

1. How many people can be accommodated in a three bedroom house?
2. Are the rates quoted on your website per person rates?
3. How much tax would be added to the daily rate?
4. Is the outdoor pool heated?

I would appreciate hearing from you soon, as I hope to have my vacation plans completed by June 1.

Christine D'Alfonso

Christine D'Alfonso

2 PLACING AN ORDER

An **order** resembles a routine request in several ways. Your reader is motivated to fill your order—that's what keeps the company in business. Your job is to make it as easy as possible for your reader to find out exactly what you want, when you want it, and how you plan to pay for it. Consequently, an order follows the same direct order as a routine request. Here are the steps:

1. **Identify your purpose in the first sentence.** *Exactly*

    Examples

    I would like to order a copy of *Microwave Mania* by Arlene McKean.

    Please send me the following items from your spring catalogue:

    If you are ordering something from an advertisement, it is helpful if you mention the name and date of the magazine or newspaper in the first sentence.

    Example

    I would like to order the Ultra Glide ski wax kit advertised in the October issue of *Canadian Skier*.

2. **Add information to identify exactly what you are ordering.** Study the catalogue, advertisement, or other source to find all the identifying information—for example, page number, catalogue or item number, and the name of the product exactly as it is written. Identify any choices you have to make about size, colour, and so on.

3. **Organize the details for easy reading.** Look at a catalogue order form for an example of how to organize information in a clear, easy-to-read format. A table format is a good choice, especially if you are ordering several items. If you are asking your reader to make a decision, make this very clear in the text—for example, "If the Tripmaster hiking boot is not available in half sizes, please send size 9, along with the Tripmaster innersole, catalogue 19 54806, size 9." Order fillers are expected to work quickly, so ensure that important information is noticed rather than skimmed over.

4. **Close with payment and delivery details.** State in your letter if a cheque or money order is enclosed. If you are paying by credit card or on an account, remember to include the account number and the expiry date and type of card (if applicable). Sign the letter with your signature as it appears on your card. You may also request payment on delivery (C.O.D.) if the catalogue or advertisement indicates that this is an option.

If you have an account with your reader, your goods will be delivered to the address stored in the computer unless you clearly request other arrangements. If you are a new customer, delivery will be to the return address on your letter—again, unless you request an alternative. If you must have your order by a certain date, mention this in your letter. Omit closing formulas like "Thank you in advance" and other tired clichés.

Read the advertisement in Figure 4.3. Using the Full Block format, use the lines following it to order two items. Compare your letter with the one in Figure 4.4.

FIGURE 4.3 Advertisement—Life's a Beach, Inc.

Born to Bronze
beach towel—thick, thirsty terry.
24" x 72"
White/bronze or bronze/white.
$18.99

No Ray Sunglasses
Eliminate harmful UVA and UVB, and blue light.
Finest polarized lenses. Brown or grey.
Frames in white, blue, red, black, or chrome.
$29.99

Authentic Panama Hat
Imported from Ecuador. Lightweight, fashionable protection in natural fibre.
White, cream, or khaki.
Please specify hat size when ordering.
$48.00

Life's a Beach, Inc.
2365 Marine Drive North Vancouver BC V7J 4T2

We pay GST. We pay shipping. B.C. residents add 8% sales tax.
CanExpress next-day delivery $5.00 extra for each order.

FIGURE 4.4   Letter to Life's a Beach, Inc.

> 39 Woodlawn Circle
> Sydney NS B2J 2E7
> June 10, 2007
>
> Life's a Beach, Inc.
> 2365 Marine Drive
> North Vancouver BC V7J 4T2
>
> Please send me the following items from your advertisement in the <u>Courier-Loyalist</u>, June 9, 2007:
>
> 1 pair of No-Ray sunglasses, brown lenses, chrome frames
> 1 panama hat, khaki, size 6 1/4
>
> I enclose a money order for $77.99. Please send by parcel post.
>
> *Nadine Klein*
>
> Nadine Klein

3 ROUTINE REPLIES

A reply is considered "routine" if you can supply the information or product requested. Otherwise the reply is "bad news." These messages are discussed in Chapter 6. Replies are easier to write than requests because contact with your audience has already been established. The message you are answering should give you important clues about the person you are writing to. Study it carefully before you plan your letter. Try to establish the age, occupational status, and educational level of the writer. The paper, the quality of the typing or handwriting, and the spelling and sentence structure, as well as the message itself, provide clues. Next, identify what the writer wants and what questions, if any, you will have to answer. Follow these guidelines in writing your reply:

1. **Open with a reader-oriented statement.** You may wish to begin directly with the answer to the first question.

Examples

QUESTION
Does the Voyage Inn in Owen Sound ON have a discount rate for seniors?

OPENING SENTENCE
Yes, seniors' discounts are available at all Voyage Inn locations.

QUESTION
Does Lambkin Knitwear make natural fibre clothing?

OPENING SENTENCE
You will find a complete selection of natural fibre knitwear on our website.

QUESTION
Can you help us authenticate this certificate? We cannot find any contact information for the institution.

OPENING SENTENCE
The certificate you enclosed was issued by the Acme Training College, which was taken over by Zenith Institute in 1986.

If this seems too abrupt, you may prefer a brief introduction.

Example

QUESTION
Can you give me a detailed description of the fibre content of the items listed below? I do not wear clothing made with any animal or insect by-products.

OPENING SENTENCE
Thank you for your interest in Lambkin Knitwear.

Remember to use the words "you" and "your" in any introductory formula. Avoid statements like "We at the Athlete's Foot Foundation were happy to receive your letter." Your reader is not interested in your emotional state.

Do not repeat information from the letter you are replying to.

✗ POOR

Thank you for your inquiry about the biofeedback program for allergy sufferers which you heard about during the September 24 edition of "Radio Noon" when Les Ragweed was describing recent results with this technology.

✓ IMPROVED

Here is the information you requested about treating allergies with biofeedback.

2. **Respond to all of the questions in the original message.** *Answer in order asked* Number your answers if the original questions were numbered. Even if you do not have the answers to some questions, or if the answer to one question makes an answer to the next one redundant, it is safer to acknowledge each question so that your reader knows that you have seen it. Answer questions in the order in which they were asked. Look at the message again to see if there are any questions that are implied without being asked directly.

Example

I was wondering if your store handled online or telephone orders, as I have two small children and I am not able to get out to shop very frequently.

A good answer to this question would mention your child-care and play area and other services for people with young children as well as provide information about shopping online and by phone.

3. **Do unto others as you would have them do unto you.** Do not draw attention to the failings of the message you are replying to. Do not feel that you can commit the same faults. In fact, if your correspondent is confused, you must be especially clear; if he or she is rude and abrupt, you must be especially polite and gracious.

4. **Use sales-oriented language** if you are writing about a product or service supplied by your organization. Compare these examples:

✗ POOR

The Belvedere Motel can accommodate your group from August 9 to 12.

✓ IMPROVED

Your group can enjoy the first-rate accommodation and conference facilities provided by the Belvedere from August 9 to 12.

5. Use a goodwill-building close. Here are some examples:

> You can enjoy the luxury of Fleecetyme mattress covers at special prices during our upcoming January "White Sale" promotion. Watch for it at your local department store.

> I hope this information assists you in planning your home security system. Please write again if you have more questions.

> *Avoid the formula "Please do not hesitate to contact me if you have any further questions." This has become an empty cliché.*

Figure 4.5 shows an effective answer to the request on page 85.

ESL TIP

A computer template is a dummy document that serves as a pattern for the layout of a real document. You can create a kind of template routine reply to help you answer written questions more quickly. Begin with a very simple all-purpose opening sentence:

Here are the answers to your questions:

Then answer the questions. If you are answering routine questions from a coworker, you can end the message with the answer to the last question. If you are writing to someone outside your organization, use a simple closing such as

You can find more information on our website at www.—.

or

You can find more information in the attached brochure.

or

Please write again or call xxx-xxxx if you have more questions.

Keep this model in a file to help you compose routine replies quickly.

FIGURE 4.5 Letter from Sea Breeze Vacation Homes

Sea Breeze Vacation Homes
235 Vista Blvd. N.   Orlando FL   34746   (407) 555-2324

www.seabreezeorlando.com

May 11, 2007

Ms. Christine D'Alfonso
1233 63rd Ave.
Edmonton AB T6J 2E5
Canada

Dear Ms. D'Alfonso

Thank you for your interest in Sea Breeze Vacation Homes. Here are the answers to your questions:

1. A three bedroom house can sleep up to eight guests in two double-bedded rooms, one twin-bedded room, and a pull-out couch in the living room.

2. All rates quoted on the website are per night for up to eight guests in a three bedroom house.

3. Florida sales tax of 6%, county lodging tax of 3%, and tourist impact tax of 1% are added to the daily rate.

4. Pool heat is not included in the daily rate but is available for an additional fee. If you are planning to vacation in June, July, or August you will find the pool stays at approximately 72°F without additional heat.

You will find a brochure enclosed with more details of the attractive and convenient homes available for your Orlando vacation, as well as information on the many recreation possibilities in the area. If you would like to reserve a Sea

(continued)

Breeze Vacation Home, please call toll-free at 1-800-555-3838. You can also reserve securely on our website.

Sincerely

Frank O'Brien

Frank O'Brien, Manager

BUSINESS CORRESPONDENCE EVALUATION SHEET

A. Visual Appeal
 1. Does the letter have a pleasing appearance (well centred, adequate margins, good paragraph length)?
 2. Are there any sloppy corrections (obvious erasures, smudges, strikeovers)?

B. Form
 1. Are all essential elements included and conventionally presented (letterhead or return address, date, inside address, salutation, complimentary closing, signature)?
 2. Have abbreviations been avoided (except for those specified by Canada Post)?
 3. Has the format been followed consistently?

C. Salutation and Complimentary Closing
 1. Are they appropriate and compatible?
 2. Are they correctly punctuated?

D. Opening
 1. Does the opening clearly establish the purpose of the letter?
 2. Does it establish a positive tone?

> E. **Closing**
>   1. Is the closing positive and calculated to leave the reader with a good impression?
>   2. Is it action-oriented, if that is appropriate to the situation?
>
> F. **Content**
>   1. Is the information that is given complete (are any essential details missing)?
>   2. Are there any irrelevant details?
>   3. Is the content presented in a logical order?
>   4. Are the paragraph divisions logical?

4 ORAL COMMUNICATION

Making Routine Requests in Person or by Telephone

If the information you need is lengthy, detailed, or complicated, a written message is your best choice. Writing saves you time and money, because the receiver gathers and verifies information on his or her time, not yours. But if your requests are simple—"When does the subway start running on Sunday?"—or you need the information right away—"How do you treat a rattlesnake bite?"—then you will probably want to speak to someone personally or over the phone. Here is a typical telephone inquiry.

CASE 1

Texmore Ltd.:	Good morning, Texmore.
Sandra:	Hello. I'd like to speak to someone who could give me some information about the sales representative's job that was advertised in this morning's *Sun*.
Texmore:	One moment please, I'll put you through to Personnel.
Personnel:	Good morning, Personnel.
Sandra:	Good morning, this is Sandra Irving. I'm calling to find out some information about the sales representative's job in this morning's *Sun*.
Personnel:	Certainly—how can I help you?

Sandra:	Will this position be based in Ottawa, or will it require travelling?
Personnel:	It will be primarily in the Ottawa area, but there will probably be some travelling required—about two or three days a month.
Sandra:	Do you provide a company car?
Personnel:	Yes, we do.
Sandra:	And is a training program provided by the company?
Personnel:	Yes, sales reps have a five-week program on full salary from the company.
Sandra:	I think those are all the questions I had. Thank you for your help. I'll be putting my résumé in the mail today.
Personnel:	Goodbye.
Sandra:	Goodbye.

Sandra followed a few simple guidelines to adapt good inquiry skills to the telephone.

1. **Make the purpose of your call clear immediately.** This will help your listener connect you to the person who can help you.
2. **Once you have reached the appropriate person, identify yourself.** This will reinforce the fact that communication is a process in which both of you are participating. In stores or offices, a name tag or sign on a desk or door may give you the name of the person with whom you are speaking. Try to use it.
3. **Ask concise, specific questions.** Open-ended, general questions put your audience on the spot, without much time to consider the best response. Breaking your request down into specific questions will ensure more reliable answers. Long, rambling questions will confuse your listener and tempt him or her to stop paying attention.
4. **Be courteous.** Personal contact requires considerate touches like "please" and "thank you," even for routine business.

Placing an Order

Placing an order over the telephone enables you to find out more about the product or service you're ordering and make appropriate decisions based on that information. On the other hand, the amount of information you must give an order taker is usually large. You will probably need to write a lot of it down and simply read it. Remember to have with you:

1. a copy of the catalogue, advertisement, or other document, open at the right page
2. the order number and name exactly as printed
3. a choice of colour, size, and anything else that must be decided on by the person ordering; make a second choice just in case
4. the shipping address, including postal code
5. a decision on the shipping method: post office, parcel service, and so on
6. the billing information, account number, expiry date, and so on

Taking Telephone Inquiries or Orders

Like a written inquiry or order, a telephone inquiry or order is an important opportunity to build goodwill with potential clients or customers. A successful organization handles every telephone call like a sales call. Look at the following case:

CASE 2

Bruce had just started working at Steel Master Office Supplies. There had been so much to do during the first few days that he hardly had time to sit at his new desk. Finally, he was enjoying a few peaceful minutes, when the phone rang.

Bruce: Hello?

Pamela Reed: Hello, is this Steel Master Office Supplies?

Bruce: Yes, it is.

Pamela Reed: Well, I'm Pamela Reed from Shamrock Insurance, and I'm calling about the office system that was featured in this month's *Canadian Business*.

Bruce: Um, I'm afraid I don't know which system you mean, Mrs. Reed — would you have the name there?

Pamela Reed: It's Ms. Reed, and I think it was called the Landmark II.

Bruce: Oh yes, uh, that's a new line, I think. What did you want to know?

Pamela Reed: We're expanding here, putting in some new offices for our senior people, and we're looking for some new ideas, something different.

Bruce: That is our latest line, pretty up-to-date.

Pamela Reed: Well, I wondered if you had some brochures or catalogues with the different components.

Bruce: Oh yeah, I'm sure we can get one of those out to you.

Pamela Reed: Well, could you send it to me at Shamrock Insurance, 4430 38th Avenue N.W., Calgary T4N 6G8?

Bruce: Sure, just let me get a pen here. What was that again?

Pamela Reed: Shamrock Insurance, 4430 38th Avenue N.W., Calgary T4N 6G8.

Bruce: Well, we'll get that right out to you, Mrs. Reed. Bye now.

Shamrock Insurance subsequently had its entire office redone—by a rival firm. How did Bruce blow this opportunity? List eight things Bruce did, or didn't do, that cost Steel Master Office Supplies a chance for a big sale. Compare your list with the one on page 99.

1. _____
2. _____
3. _____
4. _____
5. _____
6. _____
7. _____
8. _____

The following guidelines will help you handle telephone requests and orders effectively:

1. Answer the telephone by identifying yourself and your business or organization.
2. Listen to the speaker's statement of his or her purpose in calling. Determine whether you are the appropriate person to take the call.
3. Ask questions to clarify exactly what the caller wants. Do not point out that he or she is confused, inaccurate, or unclear.
4. Be alert for sales or service opportunities. Listen for information about your caller's wants and needs. Suggest ways in which your organization could meet those needs.

5. Give the caller positive feedback about his or her choices and decisions. For example, a hotel reservations clerk might affirm a client's choice by saying, "The weekend package is really a great bargain," or, "You'll appreciate the extra space in your deluxe room." This is called "resale," and it is an important goodwill-builder.
6. Confirm all details if the caller is placing an order.
7. Suggest appropriate follow-up to an inquiry: sales call, visit by caller, information in the mail, call back. Take the initiative.
8. Thank the caller. Repeat your name and indicate that you will be happy to help him or her again.

Answers to Chapter Questions

Page 98:
1. Did not give proper identification when he answered the phone.
2. Was not familiar with current promotion or products.
3. Assumed caller wished to be called "Mrs."—repeated this error after he had been corrected.
4. Failed to use sales-oriented language when discussing company products.
5. Did not look for sales opportunity when caller described needs.
6. Did not volunteer to provide more information—the caller had to ask for brochures.
7. Was not prepared to take caller's name and address.
8. Ended call without determining if the caller needed anything more.

CHAPTER REVIEW/SELF-TEST

1. Use the **direct order** to make a request when
    (a) you have the power to make the receiver do what you tell them.
    (b) you are pressed for time.
    <u>(c)</u> the request is part of the receiver's normal job responsibilities.
    (d) the request is easy for the receiver to carry out.

2. Begin a routine request by
    (a) making friendly, ice-breaking statements.
    <u>(b)</u> stating the purpose of the message.
    (c) giving the receiver some background information about who you are.
    (d) explaining why you need information or help.

3. Include details that
   (a) give the reader a sense of you as a whole person.
   (b) explain why you are making this request to this particular person or organization.
   (c) keep the audience from losing interest.
   _(d)_ will help you get a more useful response.

4. Close the message by
   _(a)_ asking the receiver to respond to your message.
   (b) thanking the receiver for taking the time to read or listen to your message.
   (c) explaining why you need information or help.
   (d) saying, "Please do not hesitate to contact me if you require further assistance."

5. When placing an order, use
   (a) any format you like. The receiver is lucky to have your business.
   (b) the "special request" format to persuade the receiver to do you the favour of selling you something.
   _(c)_ the direct order format to avoid mistakes and delays.

6. Use the **direct order** to send a reply when
   (a) you must say no to a request.
   _(b)_ you can say yes to a routine request.
   (c) the receiver has no power to object.
   (d) the receiver is pressed for time.

7. Begin a reply by
   (a) expressing your pleasure at having the opportunity to give information or help.
   (b) telling the receiver something about your product or organization.
   (c) repeating the key points from the sender's message.
   _(d)_ none of the above.

8. Help your audience by
   _(a)_ answering all his or her questions in the order they were asked.
   (b) pointing out when he or she is mistaken or confused.
   (c) answering his or her questions in the order they should have been asked.
   (d) ignoring pointless or repetitive questions.

9. When you are answering questions about a product or service provided by your organization you should
   (a) point out that this information was already available elsewhere.
   (b) keep your answers brief and factual.
   _(c)_ treat this as a sales opportunity.
   (d) all of the above.

10. End your reply with
    (a) a friendly expression and a suggestion for follow-up where appropriate.
    (b) a statement of how much you enjoyed the opportunity to give information or help.
    (c) the phrase "Please do not hesitate to contact me if I can be of further assistance."
    (d) none of the above.

Answers

1.C 2.B 3.D 4.A 5.C 6.B 7.D 8.A 9.C 10.A

EXERCISES

1. Find an ad for a product or service in which you are interested. Write a letter requesting at least four items of information not given in the ad. Use one of the formats illustrated in this textbook. Hand in **two** typed copies of your letter and a stamped envelope addressed to the company. Include the ad with your assignment.

2. Write a short (200-word) critique of the reply you receive to the letter of inquiry, above. Use the Business Correspondence Evaluation Sheet on page 94 to help you.

3. You wish to sell crests and pins with the emblem of your organization. Prepare a fax to be sent to Impact Manufacturing Ltd., 1-800-555-3784, outlining your requirements and requesting an estimate. Include a line drawing of your emblem.

4. You are opening your own business and want a distinctive letterhead. Find on the Internet the name of a company that designs stationery, and write a message outlining your requirements.

5. You are employed in the personnel department of a large retail business. In a trade journal you have seen an advertisement for *Go-to-Sell,* a training video for salespeople. The ad mentioned only one rate: "Rental as low as $39." If this video is appropriate, you would like to show it to your new salesclerks. Write a letter requesting all appropriate information, addressed to Western Media, 1948 29th St. N.W., Calgary, AB T4N 3R6.

6. Write a letter of inquiry based on the following: You have heard that the provincial government is starting a program awarding grants to companies that can provide jobs to unemployed people between the ages of 18 and 24. Write to the Ministry of Labour requesting more information. Specifically, you want to know what your company would have to do to qualify and how the program works. Consider giving some details about the nature of your company.

7. You have volunteered to organize a graduation banquet for your program. Design an e-mail message to be sent to hotels and banquet halls in your community, requesting information about their facilities. Outline your requirements, and ask at least four relevant questions.

8. Identify an organization for which you think you might like to work. Phone the organization to arrange a meeting to gather information about job opportunities for community college graduates. This is not a job interview. In your phone call, be sure to:
   (a) explain the purpose of your proposed meeting;
   (b) obtain the name of the appropriate person to meet with;
   (c) arrange a mutually convenient date and time.
   This meeting will form the basis for the research interview assignment described on page 245. The information you obtain would make a worthwhile class presentation.

9. Prepare a reply to the following e-mail message:

> **Subject:** New account information
> **Date:** Aug/23/2007 1:04:24 PM Eastern Daylight Time
> **From:** carla.mycek@hotmail.com
> **To:** municipalcreditu.ca
>
> I will be moving to Windsor in September to attend college and I would like to open an account with you. Please answer the following questions:
>
> 1. Do I need to withdraw all the money from my existing bank account and bring it with me or can it be transferred some other way?
> 2. What identification will I need to open an account?
> 3. How soon after opening my account will I receive a chequebook?
> 4. Do you offer debit cards?
>
> Please reply as soon as possible so I can open my account before school starts.
>
> Carla Mycek

10. Write a reply to one of the following letters:
   (a)

> Leisure and Recreation Department
> Bristol Local Council
> Bristol, England
> January 16, 2008
>
> Centennial Community Centre
> Arkona ON L9C 1E4
>
> I am the director of the Leisure and Recreation Department of Bristol Local Council, Bristol, England. In late May, I and a group of my staff will be touring recreational sites in southern Ontario, and we are very interested in visiting your facility, which has been recommended to us as a model of its type. We are as yet uncommitted for the 21, 22, and 23 of May, 2008. Would it be possible to arrange a tour for seven people on one of these dates? If so, how long would it last? What areas of your operation would you recommend to our particular attention? We hope to have a firm agenda by February 8, so we would appreciate receiving your reply by the end of this month.
>
> *Leslie Henderson*
>
> Leslie Henderson

(b)

Business Division, Atlantic College
1790 Tacoma Drive
Dartmouth NS B2W 6E2
April 11, 2008

Canadian Magazine Association
401 Richmond Street West
Toronto ON M5V 3A8

The Business Division at Atlantic College has recently created a budget for magazine subscriptions. Your answers to the following questions would be appreciated:

1. What business periodicals are currently published in Canada?
2. Do any of these periodicals focus on Atlantic Canada?
3. Can you recommend three or four Canadian general-interest magazines that college students might enjoy?
4. Are there special subscription rates for institutions?

Your prompt reply will enable us to order magazines in time for the beginning of the fall semester.

Kelly Fraser

Kelly Fraser

(c)

82 Maple Gardens
Godfrey ON K0H 1L2
May 2, 2009

[Your city] Tourist Information Centre
683 King St.
[Your city]

Please send me some information about tourist attractions in your area. I will be taking a summer course there in July and I would like to take some additional vacation time with my husband and children ages 5 and 10. What do you recommend for family accommodation: hotels or short-term apartment rental? What attractions are particularly suited for families in the area? My husband and I enjoy going to the theatre so we would also like information about local shows. We plan to stay about three weeks. Please reply before May 12 so that we can complete our travel plans.

Rose Ruiz

Rose Ruiz

(d)

75 Church St.
Saskatoon SK S4K 1P2
2008 01 08

[Name and address of your college]

Attention: [Your program] Course Director

Please answer the following questions regarding your program:

1. What high school courses would be good preparation for this program?
2. How important are communication skills to success in this program?
3. What job opportunities are open to your graduates?
4. Approximately how much money would an out-of-town student require for tuition, books, and living expenses for one academic year?

I would appreciate your reply before February 1, as I am planning to apply to college for September.

Barat Savunth

Barat Savunth

CHAPTER FIVE

GOOD NEWS: THE DIRECT APPROACH

Chapter Objectives

Successful completion of this chapter will enable you to:

1. Deliver good news effectively, orally or in writing, using the direct approach: good news, explanation, goodwill closing.

2. Give a good news response to a problem or complaint.

3. Deliver good news orally, using the direct approach.

INTRODUCTION

A routine message is one which the receiver expects and predicts. When you have a pleasant surprise for your audience, or when you are able to turn a negative situation into a positive one, then your message is good news.

1 WRITING GOOD NEWS MESSAGES

CASE 1

Thank you for entering the "Why I Like Old Mexicali Taco Sauce" Jingle Contest. We were gratified that so many people responded to our contest—over 40 000 happy users of Old Mexicali, the taco sauce that fights back. We at Mescalito Mexican Foods have been making this sauce for over 70 years, using only the finest ingredients, including deluxe chilis grown specially to give Old Mexicali the taste that just won't quit. And the same high standards are followed in making all our fine products—El Paso Taco Chips, Rio Grande Refried Beans, and South of the Border Down Mexico Way Chili con Carne. So you can be proud to hear that your jingle has won the first prize of $50 000. Congratulations.

Everyone would like to win a contest with a $50 000 first prize. But the lucky winner of this jingle contest may have to read about it in the paper. This letter takes so long to deliver the good news that the reader might well give up before getting there.

A message brings good news when we can reasonably assume that the audience will be immediately pleased to read or hear it. The good news may be a reply to a request, such as a job application or a customer complaint, or it may be an announcement of some favourable event. In any case, you will want to capitalize on your audience's positive response to the good news by putting it right at the beginning. This **direct order** is similar to the order you used in routine messages in Chapter 4:

1. Good news
2. Explanation
3. Goodwill close

1. Putting the **good news** first accomplishes several things. It gets the audience's attention and ensures that they receive the important part of the message, even if they do not read it all the way through. Ideally, the positive impact of the good news will motivate your audience to keep on reading.

In a memo or e-mail message, put the good news in the subject line if you can. If you do not have space, at least indicate that good news is coming in the body of the message.

Example

Subject: Good news on contract talks

2. The **explanation** follows up the good news with details your audience will naturally want or need. For example, if you are telling an applicant that she has been selected for a job interview, she will need to know where and when it will take place. If you are announcing a pay raise, your reader will want to know when it takes effect. If your explanation is too long, it will detract from the impact of the good news. Ask yourself what purpose each part of the explanation will serve from the reader's point of view. Leave out anything that isn't important right now. You can always follow up with additional details in a further message.

Handling the explanation in response to customer complaints requires a special approach. Item 2 on page 111 will give you more details.

3. Close the good news message with a brief, you-centred **expression of goodwill.**

Example

Date: 2009 06 04
To: Ingrid Forrest
From: Lance Binkley L.B.
Re: Vacation request

Your vacation dates have been changed to August 7–18, as you requested. Greg King has agreed to cover the August 9 meeting for you. Please make sure that he has all the audit material by July 15. Good luck with your plans.

The text below is the body of a letter sent to a student who had applied to a college program that was in high demand:

Thank you for your interest in the Computer Animation Arts program at Metropolitan College. Over the last three months we have been reviewing the portfolios of over 200 highly qualified applicants to select only 35 students for the upcoming year. Your application for the Computer Animation Arts program has been approved. Canada is taking a leading role in this exciting combination of art and technology. After graduation you will have an opportunity to work in advertising, television, or the feature film industry, and lifelong learning opportunities at Metropolitan College will keep your skills current and in demand. Congratulations. We look forward to seeing you this fall. Registration will take place at the main campus on September 3, 2008, from 9 AM to 12 PM. An information package will be sent to you in early August.

Beside the right number below, copy: 1. the good news, 2. the explanation, 3. the goodwill close. Omit any unneeded information.

1. _____
   _____

2. _____
   _____

3. _____
   _____

Why is this message more effective than the original letter?

2 RESPONDING TO COMPLAINTS OR PROBLEMS

Handling customer complaints requires special attention. Your audience is angry or disappointed. Your organization, and perhaps you personally, are being criticized. Even if you are prepared to make the adjustment that the customer requests, this good news is tempered by the bad news that your company or your product failed in some way the first time around.

A good news message tells customers that their problem has been looked into and that their request has been granted. It may even tell customers that they will be receiving additional compensation for any loss or inconvenience. Whenever possible, a good news message also rebuilds customer confidence by explaining what steps you are taking to prevent problems from recurring. Following a few simple guidelines for modifying the **direct order** will ensure the most effective presentation of your message.

   1. Put the good news first. The good news is your quickest means of restoring good feeling between you and your customer. The rest of your message will be bathed in its glow. Tell the reader right away that her refund is enclosed,

No Negative - all positive

her deck is being repaired free of charge, her video-phone is being delivered by courier. Leave apologies or other negative statements for later. Avoid limp, ambiguous openings such as:

Thank you for your letter describing your experiences with our Gel Flite running shoes.

or

We were sorry to receive your letter regarding the cracks in your driveway.

The following openings give the audience an immediate, positive message:

Enclosed you will find a full refund for the shoes you recently returned.

A repair person will be coming to your house early next week to repair your driveway at no cost to you.

2. **Give the customer an explanation.** Restoring customer confidence usually requires more than simply offering compensation. The customer has been disappointed or inconvenienced in some way by your company. He or she wants to be sure that it won't happen again. A good explanation conveys honest concern and a commitment to quality.

Examples

✗ POOR
We process thousands of requests every day. Inevitably some get misplaced or misfiled.

This writer implies that the company is not really concerned with providing good service. The customer will probably feel that his problem was not taken very seriously.

✓ IMPROVED
Our aim is to fill every request as promptly and accurately as possible. Although no system can be 100 percent perfect, we are always working on ways to improve ours.

✗ POOR
Exposure to temperatures above 30 degrees Celsius causes cocoa butter to rise to the surface of the chocolate and crystallize, turning the chocolate pale brown or white. This does not affect taste or quality. Nevertheless, we are prepared to exchange the chocolates you bought, if you can provide proof of purchase.

This explanation is defensive and grudging. The customer will probably feel that she is being offered compensation just to spare the company further trouble, not because there is any merit in her request. This will not restore her goodwill toward the company.

✓ IMPROVED
Exposure to temperatures above 30 degrees Celsius causes cocoa butter to rise to the surface of the chocolate and crystallize, turning the chocolate pale brown or white. Although this does not affect taste or quality, we can understand that you want your Black Beauty chocolates to look as good as they taste. Please send your proof of purchase to the address above and you will receive your free replacement box.

✗ POOR
The clerk who handled your original order didn't take the time to confirm that the parts he was dispatching were compatible with your machine.

The writer is attempting to make the company as a whole look better by passing blame to a single individual. But customers justifiably assume that a company must take responsibility for the actions of people it has hired and trained.

✓ IMPROVED
Unfortunately, the parts dispatched to you had not been matched to the specifications of your machine.

✗ POOR
It appears that poor maintenance led to a breakdown in the machine that date-stamps our orders. Everything for the last month has been dated June 14, and consequently the boys in the warehouse have just been picking them at random instead of on a first-come, first-served basis.

This explanation puts the writer's company in an unflattering, unprofessional light. The customer's image of the company will probably be worse, not better, after reading this.

✓ IMPROVED
A malfunction in the machine that date-stamps orders led to a breakdown in our priority shipping system.

A good explanation, like the following, is tactful without being evasive. It creates a concerned, responsible image for the company:

Normally, Parti-Tyme Cakes are sprayed lightly with an edible vegetable gum after icing to preserve their fresh taste and eye appeal. A jammed timing mechanism permitted one batch of cakes to go through the spraying process several times, which produced the crusty texture you mentioned. As the appearance of the cakes was not affected, several passed visual inspection and were packaged before a spot check discovered the malfunction and corrected it.

At the end of the explanation you may wish to add an apology, especially if the customer had a particularly negative experience or seems very angry. **Keep it brief,** however; don't go into details that will revive the customer's sense of deep grievance.

✗ POOR
We are terribly sorry for all the inconvenience you suffered because of the complete collapse of your balcony.

✓ IMPROVED
We are sorry for the inconvenience this caused you.

3. **Close with a positive message.** The end of the good news message should sustain the positive note of the opening. Never apologize again at the end. Avoid "don't hate us" closings like:

We hope you will consider giving Tasti-Lite diet fudge another try.

"We hope you will ... but you probably won't" is the message this sends to the customer. Instead, look forward to the re-establishment of a good business relationship with your customer — to your benefit and the customer's. Here are some good examples that are you-centred and use resale tactfully:

With your shipment we have included ten litres of our new Pumpkin Ripple ice cream free of charge — just in time for Hallowe'en. We're sure your customers will enjoy it.

Your battery operated lint-picker is now ready to provide years of trouble-free service.

Enclosed is a brochure outlining many new courses and family activities offered this fall.

*Don't end with an apology

CASE 2

"Dear Jay-Mor Realty," the letter began, "On a number of occasions I have called the attention of your Clearwater Avenue parking lot attendant to the fact that the paving needs repair in several places, creating dangerous ridges and potholes. Last Tuesday evening, as I was picking my way across the lot, I slipped on a patch of slush and fell into a puddle which had collected in one of these potholes. If I had been carrying something, I might have been seriously injured; as it was, I merely tore the knee out of a pair of good pants and drenched my suede coat with muddy water. You will find enclosed a bill for $65.97 for replacing the former and $57.63 for cleaning the latter. I would appreciate reimbursement for these expenses incurred because of your inadequate maintenance."

Well, the guy had a point. Fred Morrone, the maintenance manager, had noticed how badly the pavement needed repair. The construction strike in the fall had set the regular maintenance schedule back at least six weeks, and now the temperatures were too low to consider any resurfacing. Maybe the worst hazards could be fenced off till the spring. In the meantime, he'd better do something about this claim. Fred started drafting a reply:

> Dear Ed Drenna
>
> We were very sorry to receive your complaint about our parking lot. Due to circumstances beyond our control, we were not able to bring the parking lot surface up to standard before the winter. Hopefully this situation will resolve itself by next spring. We are sending you a cheque for the damages you claimed were caused by a fall in the parking lot. Once again, please accept our apologies.

Fred's first draft is not an effective good news reply. List seven things that make this letter a failure.

1. _____

2. _____

3. _____

4. _____

5. _____

6. _____

7. _____

Compare your list with the one on page 120.

Write an improved version of Fred's letter. Compare your version with the one in Figure 5.1.

The message in Figure 5.2 presents a problem that requires a different kind of solution. The writer does not want compensation; the person wants to know that the problem will not occur again. You will find the answer on page 118.

> ### ESL TIP
>
> When you give good news in response to a problem or complaint your main objective is to restore the goodwill of your client, customer, or employee. To achieve this objective you must ensure that the tone of your message is as positive as the good news you are giving. You can do this in several ways. First, emphasize the words "you" and "your." "Here is your refund," is a better choice than "I am going to give you a refund." You do not wish to call attention to your power as the decision-maker or your self-interest in making the customer happy. Do not use expressions like "You claim it was broken when you opened the package," or "You say you did not have this near anything magnetic." These forms imply that you do not necessarily believe the other person's statement. Avoid suggesting that fixing a problem is a chore for you. "I will send this back to the warehouse and then a new part can be attached for you," is a better choice than "I'm going to have to send this back to the warehouse and get a new part attached."

FIGURE 5.1 Good News Response to a Complaint

Jay-Mor Realty, Ltd.

948 Sherwood Ave. Winnipeg MB   R2M 6A5 (204) 555-7222

February 16, 2008

Ed Drenna
43 River St.
Winnipeg MB R3T 2P4

Dear Ed Drenna

Enclosed you will find a cheque for $123.60, as you requested in your letter of February 9.

Normally, our parking lots are resurfaced every other autumn. Unfortunately, a construction strike put our contractor seriously behind schedule, and we were unable to arrange a date for repairs to our Clearwater Avenue lot before low temperatures forced us to postpone the work until spring. We are sorry for the inconvenience this has caused you. Our maintenance staff will be systematically checking the parking lot surface next week and putting barriers around any potential hazards. Full resurfacing will take place as soon as warmer weather arrives.

Your safety and satisfaction are important to our parking lot staff. The enclosed card will entitle you to park free at any Jay-Mor lot for the next thirty days. Just have it validated by the attendant the next time you park at Jay-Mor.

Sincerely

Fred Morrone

Fred Morrone
Maintenance Manager

[Handwritten annotations: "Good news", "Explain situation", "Goodwill Closing"]

FIGURE 5.2 Message of Complaint

Subj: Washroom access for disabled patrons
Date: Oct/27/2009 1:04:24 PM Eastern Standard Time
From: Quoc Vinh Tran qvt@hotmail.com
To: www.harbour55.com

Attention: Manager

Last Friday night I visited your restaurant with some friends to celebrate a birthday. Because one of the group uses a wheelchair, I looked in the Downtown Restaurant Guide to find an appropriate place for our party.

The listing for Harbour 55 indicated that it was wheelchair accessible. I also looked on your website and saw a "disabled" icon, so I confidently made a reservation. But while we were happy to see that the entrance had a ramp and the restaurant had no interior steps, we were very unhappy to discover that the washrooms were in the basement.

It doesn't seem very realistic to call a restaurant accessible if it is not possible for someone in a wheelchair to visit the washroom. I think you should change your listing and the information on your website.

Quoc Vinh Tran

FIGURE 5.3   Good News Response to a Complaint

> Subj: Re: Washroom access for disabled patrons
> Date: Oct/27/2009 4:18:04 PM Eastern Standard Time
> To: Quoc Vinh Tran qvt@hotmail.com
> From: Alena Makavets, Manager, Harbour 55 www.harbour55.com
>
> You are right to assume that "wheelchair accessible" should refer to all the facilities of our restaurant.
>
> When the Downtown Restaurant Guide was published last January we had a main floor washroom in place for patrons who had difficulty with stairs. However, during an inspection in June we were told that the placement of the washroom was in violation of the fire code and could no longer be used. I am sorry that this caused a problem when you visited the restaurant.
>
> We have drawn up a new floor plan and are waiting for approval by the building inspectors. We hope to have a new accessible washroom by the new year. In the meantime we will take the icon off our website and add a note explaining the current situation. Please accept our invitation to visit again when the renovations are complete. You will receive an announcement by e-mail.
>
> Alena Makavets, Manager
> Harbour 55

3 DELIVERING GOOD NEWS ORALLY

The techniques suggested for writing good news messages can easily be adapted to oral communications. Most of us enjoy the opportunity to deliver good news in person. The **direct order**—getting right to the point—comes easily when you anticipate that your audience will be happy to receive your message. Here is the direct order:

1. Good news
2. Explanation
3. Goodwill closing

   1. **Give the good news as soon as possible.** Don't leave your audience in suspense with an "I suppose you're wondering why I called you in here" opening. Limit

yourself to the briefest formula: "I have some good news for you ...," or, "This is Dr. Freeman's office; I have some good news from your test results." Then deliver the good news.
2. **Add any necessary explanation.** Give your listener a chance to ask questions. Let the needs of your audience determine how long or short the explanation should be.
3. **Wind up the conversation with an expression of goodwill.**

CASE 3

Mark Dubois had recently written some exams that would upgrade his professional classification — if he passed. They were tough exams and he had been waiting almost a month to get the results. So when he was called to the Personnel Office, he was pretty nervous. Here's how it went:

Dave: Well, come in, Mark. The results of your exams are in, I have them right here. Why don't you sit down? I guess you've been kind of nervous about them. A lot of people write these but not too many pass the first time. I remember a couple of years ago someone sat them four times before he passed. But I guess it's worth it in the end. Anyway, you did fine; passed every one.

Mark: Well, that's a relief. Could I see ...

Dave: Now, of course, you'll want to make some decisions about following up, maybe going up another level. You might like to talk to Theresa Comisso — she's done that. Of course, you have to enroll within the next two years, but I guess that's not a problem unless you're planning to transfer to another branch. We're opening up a northern office, lots of good promotion opportunities, but a little isolated, you know. How'd you feel about that? Or, of course, this opens up some possibilities right here; I was reading in the last newsletter about Fred's retirement. Now, of course, this would qualify you for his area ...

Mark: I think I'd just like to look at my results for a while.

Dave: Well, sure. I just need you to fill out a receipt for your transcript and it's all yours. Why don't you give me a call next week and we can go over your options.

Mark: Okay, then ...

Dave: Well, you're lucky, you know. A lot of the people here haven't been able to keep up with their professional training. You know Al's mother's living with him now; she's had a lot of health problems. And then Christine's husband walked out; that was a real mess. And of course it takes time away from your family. But congratulations; I'll look forward to giving you any help I can with your plans.

Mark: Thanks, I'll be in touch.

Cross out everything in Dave's conversation that detracts from the impact of his good news message.

Answers to Chapter Questions

Page 114:
1. Doesn't begin with good news.
2. Uses negative word "complaint" in reference to reader.
3. First sentence implies that the company was sorry to get the letter, not sorry about the accident.
4. Explanation doesn't explain. Cliché phrase "due to circumstances beyond our control" is vague and meaningless.
5. Cliché phrase "hopefully this situation will resolve itself" implies that the company will not take any active steps to deal with the problem.
6. Phrase "damages you claimed were caused ..." suggests that reader may be lying.
7. Closes with a negative message.

CHAPTER REVIEW/SELF-TEST

1. A message should be presented as good news when
   (a) you have nothing negative to say.
   <u>(b)</u> you have a pleasant surprise for your receiver.
   (c) you are trying to sell something.
   (d) you can say yes to a routine request.

2. Begin a good news message with
   <u>(a)</u> the good news.
   (b) an expression of goodwill.
   (c) a brief introductory formula.
   (d) any of the above.

3. In a memo or e-mail message, try to put the good news
   (a) in the closing.
   (b) in a different font.
   (c) in the subject line, if you have space for it.
   (d) in an attachment.

4. The explanation section of a good news message explains
   (a) why the news is good.
   (b) how you reached your decision.
   (c) full details of all the results of the good news.
   (d) what your audience needs to know right now to take advantage of the good news.

5. The closing of the good news message
   (a) thanks the audience for taking the time to read or listen to the message.
   (b) repeats the explanation, for emphasis.
   (c) reinforces the good news with an expression of goodwill.
   (d) should do all of the above.

6. A good news response to a problem or complaint should begin with
   (a) an apology.
   (b) a statement about how you felt when you learned of the problem.
   (c) good news.
   (d) a summary of the problem or complaint as you understand it.

7. The purpose of the explanation in a good news response to a problem or complaint is to
   (a) restore faith in your product, service, or organization.
   (b) pass blame to someone who wasn't following the rules.
   (c) avoid a lawsuit.
   (d) tell the reader or listener exactly what happened.

8. An apology in a good news letter should
   (a) describe the problem in complete detail.
   (b) come at the end of the explanation.
   (c) be repeated at the end of the message.
   (d) all of the above.

9. Close the good news response to a problem with
   (a) a promise that no problems will ever happen again.
   (b) a statement of how you feel about the inconvenience your customer or client experienced.
   (c) a hope that the audience will give your product or service another try.
   (d) none of the above.

10. Re-establish goodwill at the end of your message by
    (a) offering additional compensation to your audience.
    (b) looking forward to a continued good business relationship with your audience.
    (c) a you-centred statement.
    (d) any of the above.

Answers

1.B 2.A 3.C 4.D 5.C 6.C 7.A 8.B 9.D 10.D

EXERCISES

1. Revise the following messages to improve their tone and organization:
   (a) Thank you for your letter regarding the explosion in your VCR. We have examined the machine in our lab, and it appears that a faulty connection allowed a spark to contact gases apparently generated by the head cleaner you were using. We are glad to hear that no one was seriously injured. We will be sending you a replacement machine shortly. I hope you will continue to rely on Lectron for all your electronics needs.
   (b) We were interested to hear of your experience with our Right Off stain remover. It appears from our analysis of the blouse you sent us that the spray has combined chemically with the dimethylcellulose protein in the fabric to produce those large orange blotches when heated in the dryer. We have received several similar complaints, and our product scientists are working on changing the formulation of Right Off. All existing stock will be relabelled in the meantime. We are enclosing a cheque for $38.42, as requested, to cover the cost of replacing the blouse, as well as some samples of our other fine cleaning products. Thank you for drawing this problem to our attention.
   (c) Thank you for applying for the job of assistant retail manager. Many fine candidates were interviewed for the job, and it was indeed a difficult decision to select one person from so many qualified applicants. Therefore, we at Marks and Sparks are happy to be able to offer you this position. Please phone 481-1111 for information regarding starting date, required documentation, etc.
   (d) Memo:
   Jim, it's always difficult to change vacation schedules once they're set. People have made plans around their dates, and naturally these have priority. In addition, our secretarial and other support requirements are fixed according to the staff who will be in the office at any given time. However, it seems that Louise had her reservations in Quebec City fall through, so she is willing to trade July 8–27 for August 2–22. So—bon voyage!

2. You are the manager of Trendway Catalogue Sales. Prepare the following good news messages:
   (a) To all customers: To benefit the environment, Trendway will be eliminating over-packaging and using lighter, less bulky mailers made from recycled materials.
   (b) To Preya Ramroop, a Trendway employee: Because of her leadership in developing these packaging changes, this employee will be receiving an Excellence Award from the company.

3. Kindly Fruits of the Earth Natural Market has been named Canadian Small Business of the Year by the *Canadian Business Post*. As president of Kindly Fruits of the Earth, write an e-mail to be sent to all employees, telling them of this honour.

4. Write a good news response to one of the following messages:
   (a)

   > 5990 Kingston Rd. #238
   > Toronto ON M2N 5Y7
   > 2006 04 06
   >
   >
   > Bloomsday Florists
   > 3080 Yonge St.
   > Toronto ON M4N 3P3
   >
   > Three months ago I ordered four table centre arrangements from you for a party I gave last Saturday. When the flowers were delivered Saturday morning they seemed droopy but I assumed some water would pick them up. Instead, by party time they were losing petals and looking faded. This was very disappointing, especially as I had planned to present the flower arrangements to the guests of honour after the party. I am enclosing the bill and I would appreciate a refund for the full purchase price.
   >
   > *Moira Cole*
   >
   > Moira Cole

(b)

194 Gleneagle Ave.
Toronto ON M4K 3P2
February 28, 2008

Underwear Underworld
595 Main St.
Flesherton ON N4G 1E6

When is two days the same as two weeks? When it's the two weeks spent waiting for your company's guaranteed two-day delivery. On February 11 I faxed an order to you for three pairs of red silk briefs as a Valentine's Day gift for my husband. As your order form promised two-day delivery anywhere in Canada I was confident that the gift would be here in good time. Instead I waited until February 25—long past the special day. I think it would be appropriate to refund the price of the briefs as well as the $15 I paid for express delivery. A copy of the packing slip is enclosed.

Jing-mei Li

(c)

Dept of Earth Studies
Sanford Dettweiler Memorial College
3250 Bonaventure Park Blvd. W.
Kapuskasing ON K2J 8V8
February 28, 2009

Acme Management Institute
40 Alta Vista Dr.
Ottawa ON K0K 1G3

Enclosed you will find my application for your Supervisory Skills Workshop to be held May 1–4. I attempted to register online, as I note that you offer a $25 discount to online registrants; however, after I spent quite a bit of time filling out the form it was not accepted because my address was too long for the field. As this was not my fault I think it would be fair to give me the $25 discount anyway.

Gayathri Sivalogonathan

Gayathri Sivalogonathan

(d)

Subj: Car rental ref. #618039967
Date: Oct/19/2008 1:04:24 PM Eastern Daylight Time
From: Lincoln Barnes linbarn@sympatico.ca
To: www.eeezed.ca

Two weeks ago I booked a car with you for the Thanksgiving Day weekend. I specified a non-smoking vehicle, but the car I was given had no "No Smoking" sticker and smelled distinctly of smoke. Later I found an EEE-ZED logo air freshener in the glove compartment, as though someone had quickly taken it off the rear-view mirror and hidden it. If no non-smoking car was available I think I should have been told and given the choice of cancelling or receiving a discount or free upgrade.

Lincoln Barnes

CHAPTER SIX

BAD NEWS: THE INDIRECT APPROACH

Chapter Objectives

Successful completion of this chapter will enable you to:

1. Deliver bad news appropriately, in writing, using the indirect approach: neutral opening, explanation, bad news, and goodwill closing.

2. Deliver bad news in response to a complaint or problem.

3. Deliver bad news orally, using the indirect approach.

INTRODUCTION

If good news is a message that you expect will please your audience, bad news is anything you expect to have the opposite effect. It may be a negative answer to a request. It may be the announcement of some change for the worse—increased prices, layoffs, or reduced benefits or privileges. The biggest challenge is delivering bad news in response to a complaint or problem. In this situation, your audience is already in a negative frame of mind. The most important thing to remember is to keep your message clear and logical, avoiding anything that suggests that the news is bad because your reader or listener is a bad person. Bad news is unavoidable, but a well-organized message can help your audience deal with bad news in a productive way.

1 WRITING BAD NEWS MESSAGES

When your message is definitely "bad news," it requires tactful handling. The **indirect order** breaks it gently:

1. Neutral opening
2. Explanation
3. Bad news
4. Goodwill close

1. **A neutral opening avoids starting off on a negative note,** without going too far in the opposite direction and leading the audience to expect a positive message. A factual statement is a good way to open the subject without making it obvious that bad news is coming. Here are some examples:

   For the past three years, A Woman's Place has provided its meeting rooms free of charge to outside groups.

   I have been putting together the vacation schedule for June, July, and August.

   Revise the following subject lines to eliminate the negative tone:

   (a) To: All employees   Date: 2009 04 18
   From: Marie Turco, Human Resources MT
   Subject: Reduction in Employee Benefits

   _____

(b) To: Allan Singh   Date: July 31, 2007
    From: Don Snyder DS
    Subject: Loss of Entitlement to Company Car

(c) To: Carla Williams   Date: 2008 01 15
    From: Ray Lebeau RL
    Subject: Failure to Meet December Sales Quota

---

2. **The explanation prepares the audience for the bad news.** To do this effectively, it must be clear and logical.

   Examples

   ✗ POOR
   Bottom-line considerations require us to do something about the continuing revenue deficit in our parking lot. Car owners have enjoyed a "free ride" long enough—it's time for you to pay your fair share.

   ✓ IMPROVED
   For the last two years, parking lot revenues have not covered the cost of maintenance, snow removal, and security. Covering the deficit from the general budget means that transit users, many of whom are in lower-paid job classifications, are in effect subsidizing car drivers.

   Both explanations make it easy for the audience to predict that an announcement regarding higher parking fees is coming. But the first example does not give any specific reasons; "bottom-line considerations" sounds like management jargon for plain old greed. The second sentence passes judgment on car drivers, and its use of the word "you" makes the accusation of unfairness offensively personal. The result of this paragraph would probably be resentment and anger over increased parking fees.

   The second explanation tells the audience specifically what the problem is. It appeals to a sense of fair play without accusing anyone of deliberately ripping off anyone else. The audience is now prepared to see the increased parking fees as necessary and reasonable.

> **No Negative words**

3. **Now you're finally ready to deliver the bad news.** If your explanation is clear and complete, your audience should be expecting bad news as the logical outcome. You can emphasize this relationship between the explanation and the bad news with a phrase like "As a result ... " or "I think you can see from this...."

    Try to deliver the bad news in an impersonal way. This is not the place for an attitude that stresses the "you" in the message. Avoid words with strong negative overtones, such as "reject," "deny," or "refuse."

    Examples

    ✗ POOR
    For this reason, your request for extra vacation time has been denied.

    ✓ IMPROVED
    For this reason, extra vacation time cannot be approved.

4. **Close on a note of goodwill.** Do not end with the bad news, or with an apology. It may be possible to point out a positive feature in the "bad" news. Perhaps you can offer a helpful suggestion. At the very least, offer to provide information to anyone who wants to ask questions or make comments.

    Examples

    ✗ POOR
    We're sorry that the ventilation system won't be fully operational until June.

    ✓ IMPROVED
    By the time warm weather arrives in June, our all-new, state-of-the-art ventilation system will be fully operational.

    ✗ POOR
    Consequently, you will be unable to purchase hot drinks from the cafeteria during the week of November 13–17.

    ✓ IMPROVED
    During the week of November 13–17, you may wish to buy coffee or tea from Wilt's Variety next door, while the cafeteria equipment is being replaced.

The following is an example of a bad news message that uses the **indirect order** effectively:

May 27, 2008
To: All Employees
From: Ralph Choy, Director *RC*
         Physical Plant
Re: Changes in Food Service

Over the past four months we have been surveying the operation of our three food outlets. Lunch and coffee-break items are enjoying good sales, but the hot breakfast business has been losing customers. While the number of breakfasts sold has declined steadily, the cost of opening the grill has increased sharply. In order to maintain full breakfast service, prices would have to be increased by 48 percent, or a lesser increase would have to be subsidized by lunch and snack eaters. In preference to these alternatives, the cafeteria will be ending hot breakfast service as of June 1. A larger selection of self-serve items will be available for early morning eaters. The cafeteria staff welcomes suggestions for appropriate items. Please call me at X490 if you have any comments or concerns regarding food service.

2 RESPONDING TO COMPLAINTS OR PROBLEMS

Generally speaking, good news should be your first choice when dealing with problems and complaints. While it may cost time and money to make your customer, client, or employee satisfied, you will generally find it a worthwhile investment. But there will inevitably be times when you cannot respond positively to a complaint. The claim may not have a leg to stand on, no matter how generously it's interpreted. There may be some justification for complaint, but the compensation or change requested may be out of all proportion to the problem. Or the complaint may be valid, but you cannot make the adjustment requested—for example, when replacements are no longer available. So you can't always avoid delivering bad news.

1. **Begin with a neutral statement.** Just as the good news opening in a good news message sets a positive tone for the whole message, a bad news opening would negatively affect the reader's view of the entire contents—assuming he or she reads that far. **Never** begin with the bad news. Find something neutral and factual to say. Here are some examples:

The battery charger you recently returned to us has been thoroughly tested in our laboratory.

I have spoken to the claims representative who originally handled your accident report.

The personnel department has supplied me with a record of your sick leave and vacation days for the past eighteen months.

Referring to an area of agreement or empathy is another good way to open. For example, if a customer's message began "I booked my vacation through TraveLand Tours because your brochure promised first-rate professional service," you could begin by saying, "You are right to expect first-rate service from every TraveLand agent." These openings invite reader agreement and set a positive, reasonable tone without raising false hopes. Do not apologize in the first line — or anywhere else in a bad news message. An apology implies that your company is accepting **responsibility** for the problem.

Here are five possible opening sentences for a bad news message. Put a check mark beside the ones that achieve a neutral, factual tone. Rewrite any that are not appropriate.

(a) Your Timecraft watch has been carefully inspected by our repair department.

✓

(b) We regret to inform you that your cosmetic dental work is not covered by your dental plan.

Our dental plan covers all work required to maintain good health. Unfortunately

(c) Thank you for telling us about your problems in keeping up with the payments on your loan.

Your acct was thoroughly reviewed after following your request of a revised payment plan

(d) Offices in the accounts receivable division are assigned on the basis of job classification and seniority.

✓

(e) We were sorry to hear that you were not happy with the service you received at our parts department last Tuesday.

I have spoken to the representative who served you last Tuesday in the parts dept.

2. **Give an explanation.** The explanation is even more important in a bad news message than in a good news message. A successful explanation presents your side of the situation persuasively and prepares the reader to accept bad news as the logical consequence. The explanation should review the facts that are relevant to your decision, leading the reader clearly from step to step. Never attempt to intimidate the reader by resorting to emotional or moral judgment. This will only invite the reader to become emotional and judgmental, too. This is not a helpful climate for a professional relationship.

No blame

Examples

✗ POOR
Obviously, you failed to read the instructions.

✓ IMPROVED
In order to produce a smooth, bubble-free finish, each coat of Nu-Gleem must be allowed to dry thoroughly before another coat is applied. The package directions suggest a minimum drying time of eight hours. From the description in your letter, it appears that a second coat of Nu-Gleem was applied before the first was completely dry.

A fair decision has benefit for your audience as well as for you. You can point this out in a tactful way.

✗ POOR
Surely you don't expect to be able to return something you bought "As Is" at an end-of-season sale.

✓ IMPROVED
By shopping during our end-of-season "As Is" sale, you were able to buy your patio table for approximately half the regular price. We offer these markdowns in order to clear floor and warehouse space for the new season's merchandise. The savings in inventory and storage costs are passed on in special prices to you. These items are, however, generally one-of-a-kind and may have been on display for some time.

Do not become defensive.

✗ POOR
Our chicken pies are made from the finest government-inspected ingredients. No one has ever complained of a "funny taste" before.

✓ IMPROVED
The ingredients and seasonings used in Chicken Ranch Pot Pies are selected to appeal to the taste of typical Canadian consumers, as revealed in surveys, test groups, and comments from our customers.

3. **Deliver the bad news.** If your explanation has been clear and well planned, your reader should already be expecting bad news as a logical outcome of the facts. An introductory word or phrase like "consequently" or "for this reason" will underline the connection. The bad news should be stated in a neutral tone.

    Keep it brief but unambiguous. Hope dies hard, so make sure your audience doesn't misinterpret your overly tactful refusal. Don't make the mistakes these writers made.

   Examples

   ✗ POOR
   The delicate colours of your Airloom blouse were achieved by using pure vegetable dyes. These dyes are unstable in water; consequently, the blouse is clearly labelled "Dry Clean Only" both on the package and on the fabric care label sewn into the blouse itself. The blouse you returned to us has apparently been machine washed, causing the dye to bleed and fade. We are sorry, but we cannot replace the blouse free of charge as you requested.

   The writer's explanation is clear and logical. Then she destroys its impact by apologizing, as if the company's decision were something to be ashamed of.

   ✓ IMPROVED
   As a result, the blouse cannot be replaced free of charge.

   ✗ POOR
   Health Department regulations prohibit the resale or exchange of bathing suits. Consequently, we regret to inform you that we must reject your request for a refund.

This reply is loaded with negative words: "regret to inform you" sounds like a death announcement, and "reject" is a very hostile word. It is the customer who will probably feel rejected.

✓ IMPROVED
Consequently, we cannot offer a refund or credit for a returned bathing suit.

✗ POOR
As the malfunction is not covered by our warranty, it does not appear that a refund is justified at this time.

The phrase "at this time" leaves the reader in doubt. Will a refund be justified if he tries again?

✓ IMPROVED
For this reason, the warranty on your air conditioner is no longer in force.

✗ POOR
Obviously you are not entitled to a reduction.

The word "obviously" labels the customer as either dishonest or stupid for even asking.

✓ IMPROVED
Consequently, the amount of $85.50 is still owing on your account.

4. **Close with a positive message.** Now that the painful part is over, use the closing of the message to help the reader find a positive solution to his or her problem. If possible, offer an alternative, as in the following example:

For this reason we cannot refund the purchase price of your Lawn Girl lawn mower. However, if you would like to come in and look at the full Lawn Girl line, I think you would find several models with the features you require. Should you wish to invest in one of these, you will receive a generous trade-in allowance on your used mower.

At least assure the customer that you will be glad to explain your decision in more detail or consider anything further he or she has to say. Underline your sincerity with a sentence like the following:

If you have any questions or you would like to discuss this further, please call me at 222-1234.

Do not apologize for your decision. Leave your customer with an impression of what you **can** do for him or her, not what you can't or won't.

CASE 1

When I joined The Thin-Is-In Gym, your employee told me that your exercise program could help me lose fifteen pounds in two months. Well, I've been working out for six weeks and I haven't lost a thing. Your skinny instructor stands up there twisting herself into a pretzel, I'm killing myself trying to follow along, and at the end of it all I'm not a pound lighter. So I sit downstairs with my coffee and butter tarts and feel depressed. Well, who needs it? I want my money back.

Write a bad news letter to Darlene Winter in the space provided below. Compare your answer with Figure 6.1.

FIGURE 6.1 Bad News Response to a Complaint

THE THIN-IS-IN GYM
45 Maywood Drive Oakville ON L6P 1N9

November 4, 2007

Ms. Darlene Winter
19 Brock Ave., Apt. 602
Oakville ON L9Z 6F5

Dear Ms. Winter

The Thin-Is-In Gym <u>can</u> help you lose fifteen pounds in two months.

To achieve this goal, you need to follow a routine of exercise and diet in accordance with the personal assessment you received when you became a member. Because every person's body is unique, we do not guarantee a specific weight loss in a specific time. But following your exercise and diet plan will enable you to meet your weight loss goal at a rate that is safe and healthy for your body. For this reason, we cannot offer a refund of membership fees.

However, we would like you to come in for a personal meeting with our dietician and fitness specialist. They will help you re-evaluate your program and get back on the road to your ideal weight. As an added incentive, your membership will be extended for an additional six weeks at no extra charge.

Sincerely

Gwen Nguyen

Gwen Nguyen, Manager

CASE 2

Mario Simard, manager of Maple Leaf Video, wasn't around on Thursday night when the fight broke out, but he heard all about it from Dave, the clerk on duty. While he was checking out a couple of DVDs for a customer, Dave noticed on the screen that the customer had one that was overdue; in fact, it had been out for almost two months. When he drew this to the customer's attention, the customer denied all knowledge of this item and demanded that Dave delete it from his file. Of course, Dave wasn't authorized to do this, and a loud and pointless argument took place while the line of impatient customers got longer and longer. Finally the customer threw his DVDs at Dave and stomped out of the store. Mario had been expecting some follow-up, and today he received a letter from the customer:

> Last Thursday, while I was attempting to take out some DVDs from your store, your clerk accused me of having an overdue DVD. Since I have never even heard of this item, let alone rented it, it is obvious that your store has made a mistake and entered someone else's DVD under my number. I tried to explain this to your dimwit clerk but, of course, he was covering up for someone else's incompetence and wouldn't listen to reason. Unless you sort this out quick, you won't be seeing me or any of my kids in your store again.

You can read Mario's reply in Figure 6.2.

ESL TIP

Before you start to compose a response to someone who has a problem or a complaint, ask yourself these questions:

- What is my most important objective in sending this message?
- Do I have any other objectives?
- What do I know about my audience?
- What is my plan for achieving my objectives with this audience?

Remember that building a good relationship with customers, clients, and employees is an objective whenever you communicate. If you must deliver a negative message, you will have to plan carefully to avoid creating bad will.

FIGURE 6.2 Bad News Response to a Complaint

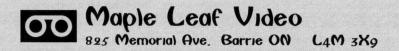
Maple Leaf Video
825 Memorial Ave. Barrie ON L4M 3X9

September 20, 2008

Mr. Kevin Aucoin
33 Georgian Dr.
Barrie ON L4T 6D2

Dear Mr. Aucoin

Your membership file for the last three months has been carefully checked by our staff.

When you borrow an item from Maple Leaf Video, an automatic scanning device reads the bar code on your membership card and on the items you have selected. It is not possible for rentals to be wrongly charged to your account through clerical error, since no manual data entry takes place.

Our records show that <u>Night Call Nurses Behind Bars</u> was borrowed on your card on July 24, 2008, and has not been returned. If you cannot find this item, a $60.00 replacement fee is payable, as described in the cardholder agreement you signed when you joined Maple Leaf Video. This fee covers the purchase and cataloguing of a replacement. In addition, late fees of $3.50 a day are payable, which reflect the revenue lost while the DVD has been unavailable for rental to other customers. Normally fees stop accumulating when the replacement fee is paid, but in view of the concerns expressed in your letter I have put the late fees on hold as of September 18, provided that your account is settled by September 25.

Copies of your rental record and cardholder agreement are enclosed. If you have further questions, please call me at 555-2139.

Sincerely

Mario Simard, Manager

3. DELIVERING BAD NEWS ORALLY

The techniques suggested for writing bad news messages can easily be adapted to oral communications.

Given a choice, most of us would prefer to deliver bad news in writing. The personal element in oral communication, and the potential for immediate feedback, make it an unattractive choice. But you will often have to deliver bad news in person or over the telephone. Someone may ask you a direct question: "Did I get the job?"; "Can I change the date of my flight?" Or your audience may need the support of a personal message. A bad news message with serious consequences for the receiver, such as job loss, demotion, or a poor performance review, should always be delivered orally, although you will probably follow up with a written message. An effective bad news message uses the **indirect order** to break it to them gently:

1. Neutral opening
2. Explanation
3. Bad news
4. Goodwill closing

1. **Don't blurt out the bad news right away,** even in response to a direct question. A **neutral opening** raises the subject while ensuring that you will have a chance to explain why the news has to be bad. Don't announce that you have bad news in store.

   Examples

   ✗ POOR
   Come in, Toni. I'm afraid I have some bad news about your application for reclassification.

   ✓ IMPROVED
   Come in, Toni. I want to talk to you about your application for reclassification.

   ✗ POOR
   No, Mr. Ang, your refund cheque hasn't been processed yet.

   ✓ IMPROVED
   Well, Mr. Ang, we're processing about 500 refund requests a week here.

2. **Prepare your listener for bad news with a clear, logical explanation.** Keep your tone matter-of-fact. Don't sound as though you are apologizing or looking for approval from your listener.
3. **Deliver the bad news.** If your decision is final, be sure that your message makes this clear.

   Examples

   ✗ POOR
   I think, then, that we'll have to consider this a non-warranty repair.

   ✓ IMPROVED
   The repairs to your transmission will not be covered by the warranty.

   ✗ POOR
   I don't think that we can back-order sale items.

   ✓ IMPROVED
   Sale items cannot be back-ordered.

4. **You may pause for a minute or two while your audience digests the news.** Be prepared to go through the explanation again or to answer any other questions. These questions are usually a way of coming to terms with the bad news, not a personal reflection on you. In any event, try not to take any reaction personally. Phrases like "I understand your disappointment" or "I can see your problem" express sympathy without undermining your decision. Once you sense that your audience has dealt with the news, close the conversation with a **goodwill closing.** This could be a helpful alternative or an offer to answer any more questions. Do not apologize for your decision.

CHAPTER REVIEW/SELF-TEST

1. Begin a bad news message with
   (a) the main point of the message: the bad news.
   (b) an apology.
   (c) a statement that introduces the subject in a neutral way.
   (d) a positive statement.

2. A good explanation
   (a) is clear and logical.
   (b) avoids blaming the reader or listener.
   (c) prepares the receiver for the bad news.
   (d) all of the above.

3. The bad news should come as
   (a) the logical conclusion of the explanation.
   (b) a surprise to the receiver.
   (c) a personal, "you-centred" message.
   (d) none of the above.

4. End a bad news message with
   (a) an apology.
   (b) a helpful suggestion.
   (c) the bad news.
   (d) an upbeat statement like "Have a nice day."

5. Respond to a problem or complaint by giving bad news when
   (a) the request is rude or poorly expressed.
   (b) you feel like it. You're the boss.
   (c) you can find any loophole or technicality that would justify saying "no."
   (d) the solution that has been requested is impossible or inappropriate.

6. You may open your response with
   (a) an apology.
   (b) a reference to something you and your audience both agree on.
   (c) a statement that suggests that good news is on the way.
   (d) the bad news.

7. A good explanation
   (a) points out that the problem was caused by your reader or listener, not by you.
   (b) questions the honesty of your reader or listener.
   (c) points out that no one else has complained about this problem.
   (d) does none of the above.

8. Deliver the bad news in a way that
   (a) shows you are sorry you have to say no.
   (b) makes it clear that you are saying no.
   (c) leaves the door open to either a positive or a negative interpretation.
   (d) is "you-centred."

9. End your bad news response to a problem by
   (a) offering an alternative solution, if available.
   (b) apologizing for having to say no.
   (c) closing with the phrase "We are sorry we were not able to help you."
   (d) thanking the receiver for bringing the problem to your attention.

10. It is a good idea to deliver a bad news message in person when
    (a) you are angry or upset.
    (b) you are afraid that putting it in writing might lead to a lawsuit.
    (c) the receiver is likely to be made angry or upset by the news.
    (d) all of the above.

Answers

1.C 2.D 3.A 4.B 5.D 6.B 7.D 8.B 9.A 10.C

EXERCISES

1. Revise the following messages to improve their tone and organization:
   (a) I can see by your letter that you know very little about car maintenance. The fact that you were not aware of any problems with your car before you brought it in for a spring tune-up is irrelevant. A good mechanic is trained to spot trouble *before* a major breakdown occurs. All the parts that were replaced were just about to give out. This is to be expected in a car as old as yours. Consequently, you are not entitled to a refund. We look forward to seeing you next fall.
   (b) We regret to inform you that we cannot replace your no-wax flooring free of charge. According to the sales representative who examined it at your house, you failed to follow instructions regarding proper cleaning and maintenance. We cannot guarantee a finish that has been abused with steel wool and cleaning solvent. Thank you for choosing Acme Beautifloor. If we can be of any further assistance, please let us know.
   (c) This will acknowledge receipt of your order dated February 1. Unfortunately, we cannot fill your order as requested. The Cheese 'n' Cracker Hostess Pack is a specially prepared hamper of delicious gourmet foods; you must realize that if we allowed customers to make substitutions we would not be able to sell it for $19.95 and still make a profit. We have sold hundreds of these gift hampers and our customers have always been delighted with the selection of cheeses—we are sure that you will enjoy the peppermint cheddar once you try it. We are sorry we are unable to help you.

(d) We regret to inform you that we must deny your request for a refund. If we allowed everyone to return merchandise just because it didn't match their decor, we'd go broke in short order. Deck furniture is seasonal merchandise, and if we don't sell it by June we'll be stuck with it all winter. Thank you for shopping at Waldorf's.

(e) Thank you for writing us about your unfortunate experience with Pearlite shoe dye. However, we cannot assume responsibility for your shoes when the directions are not followed properly. When used as directed, Pearlite will not harm leather or suede. We are enclosing a coupon good for a free package of Pearlite in a colour of your choice.

2. You are the manager of The Daily Grind Gourmet Coffee Shop. Prepare the following bad news messages:
   (a) To all members of your "buy ten cups, get one free" coffee club: Because of changes in tax regulations, sales tax and GST are now payable on the "free" cup.
   (b) To all employees: Because of a dramatic increase in claims, premiums for extended health benefits will be going up 14 percent on June 30.

3. Owing to lack of use and budget constraints, the Royal Bank will be removing its banking machine from the foyer of the Acme MicroChip building. As employee relations officer at Acme, write a memo to all employees.

4. Write a bad news response to one of the following messages:
   (a)

> Subject: Swimming Lessons
> From: Kristine Vovtchok
> Date: Thurs, 28 Oct 2008 14:01-4000
> To: Ross Rebagliati Community Centre
>
> Yesterday when I dropped my daughter off in the beginner's class at 4 o'clock I stayed for a while to watch the lesson. I can't say I was impressed. To me, it looks as though the kids are just having a good time instead of learning anything. I'm not paying good money for playtime in the pool. I want to see some work out there.

(b)

3250 Lakeshore Blvd. Apt 3110
Toronto ON M9D 1E6
2004 10 07

Gemstone Financial Services
20 Carlson Court
Toronto ON M9W 6V4

When I contacted you about investing my early retirement severance package, you assured me that I would be receiving guidance from a trained professional. Yet virtually all the stocks I invested in on the advice of your company's representative have gone down. My portfolio is worth less now than it was a year ago. I think I'm entitled to a full refund of the purchase price.

Al Fresco

Al Fresco

(c)

Subject: Your rip-off wing night
From: Chris Zamora
Date: Wed, 18 Nov 2007 14:01-400
To: Riverside Sports Bar <RSB.net>

Listen, creeps. What kind of joint are you running? You advertise Tuesday as "All You Can Eat" Wing Night—eight bucks a head. So I show up yesterday to give them a try. Not bad wings, but after five, maybe six trips to the buffet, the guy starts doling them out by ones and twos. Pretty soon I'm burning more calories walking back and forth to the buffet than I'm getting from the wings. Then he tries to stop me from dipping into the dip. So is it All You Can Eat or what? Get rid of that guy or I'm calling the paper or the Department of Health or something.

(d)

54 Myrtle Gardens
New Liskeard ON P0J 1J0
November 18, 2008

Sunseeker Tours
138 Ridpath Dr.
Toronto ON M4L 2K3

I recently purchased a holiday package to the Dominican Republic from you, transaction #58-622A. Since I made this booking I have been talking to some friends who were there last year and they recommended a hotel, the Sand Dollar, which they said was much better than the Hispaniola Princess where our tour is booked in. So I would like a refund for the hotel portion of the package so I can make my own arrangements at the Sand Dollar, or you could just change my reservation to the Sand Dollar, which would be even more convenient.

Karri Pasqualini

Karri Pasqualini

(e)

Subj: Removal of my name from mailing list
Date: Feb/06/2007 1:04:24 PM Central Standard Time
From: Elaine Littlemore
To: www.cufdl.org

When I made a donation to the Canadian Urban Fowl Defence League I did not expect to be bothered by an avalanche of requests from every pigeon, sea gull, and Canada goose organization in North America. Three attempts to have my name removed from your mailing list have failed; meanwhile the list has been circulated to an ever-growing army of fundraisers. When will this wasteful invasion of my mailbox and my privacy come to an end?

Elaine Littlemore

CHAPTER SEVEN

PERSUASIVE MESSAGES

Chapter Objectives

Successful completion of this chapter will enable you to:

1. Identify the four elements of persuasion.

2. Identify the Attention, Interest, Desire, and Action elements (AIDA sequence) in a persuasive message.

3. Apply the AIDA sequence in writing an effective sales message.

4. Apply the AIDA sequence in writing a special request.

5. Apply the AIDA sequence in writing a complaint.

6. Apply the AIDA sequence to supervisory communication.

7. Adapt the AIDA sequence to oral communication situations.

INTRODUCTION

All communication is persuasive insofar as it requires the willing participation of both sender and receiver. As you learn more in this chapter about the four elements of persuasion—credibility, facts, logic, and emotion—you will recognize that you have already been applying them in routine correspondence and good and bad news messages. The distinguishing feature of the messages discussed in this chapter is that they do not simply ask the audience to agree—they also ask the audience to make a decision requiring change. The AIDA sequence helps you to lead your audience to the decision you want.

1 ELEMENTS OF PERSUASION

Four elements—credibility, facts, logic, and emotion—determine whether your audience will find your message persuasive.

Credibility

Credibility is the quality of being believable, reliable, and trustworthy. For example, credentials like degrees and diplomas, published books and articles, and appearances on television or on the lecture circuit can establish someone as an "expert" whose views are worth paying attention to. Your relationship to your audience may also enhance your credibility—for example, if you are a doctor speaking to a patient, a teacher addressing a class, or an experienced employee showing a rookie the ropes. On a more emotional level, recognition and positive feelings enhance credibility even if no real expertise is involved; this is why advertisers want TV stars to sell coffee, or baseball players to promote razor blades.

Visual impact contributes significantly to establishing (or destroying) the credibility of an unfamiliar writer or speaker. In written communication, the impact is created by the appearance of the message itself. In oral communication, your own appearance contributes strongly to your credibility.

All of these factors help establish your credibility at the beginning of a communication situation. What happens next will depend on whether your message confirms or undermines this initial impression. In the long run, credibility is the result of your audience's positive experience.

Facts

Most people would like to believe that **facts** play the most significant role in their decision-making. At election time, voters read newspapers, watch television coverage, and attend meetings to learn more about candidates and issues. Consumers read advertisements, study labels, go on the Internet, and consult ratings of products they are thinking of buying. Students check out the job market by talking to teachers, counsellors, and people already employed at a particular workplace. Facts are the backbone of decision-making. No matter how skillfully you employ other techniques of persuasion, you will not succeed in persuading your audience if you cannot supply the concrete information they require.

The key is to assemble the appropriate number of facts. A company that wishes to order a complete new record management system will require a raft of detailed cost-effectiveness information. A group of employees planning the menu for a staff party would need only very few, simple details to make appropriate choices. Chapters 12 and 13 discuss in detail facts and research for the informal report.

Logic

Logic is the process of putting together facts to arrive at conclusions. For example, if you observe that whenever you eat chocolate you develop a headache, you may conclude that there is a logical relationship between these two events, namely, cause and effect. Chapters 12 and 13 discuss logic in a more formal sense as an element of report writing. In more informal persuasive situations, the function of logic is to show your audience the significance of the facts you are presenting and help them move from the facts to the decision you desire. Unless your facts are clearly related to each other, to your audience, and to the decision you are seeking, your information will have little impact.

Emotion

As much as we like to see ourselves as informed, logical decision-makers, few of us would deny that **emotion** often plays a large role in the decisions we make. Otherwise we would buy only long-wearing, serviceable clothes in colours that never showed the dirt, eat only nourishing lunches, and date only people with well-paying jobs and spotless driving records. But human nature doesn't work that way. Whether it's junk jewellery, junk food, or junk relationships, we all make choices that can't be justified by facts or logic.

Effective persuasion recognizes the role of emotion in decision-making and exploits it **to an appropriate degree.** A formal business proposal would concentrate on concrete specifications and the "bottom line"; it would include a subtle emotional appeal to prestige and esteem as an added incentive to a decision primarily based on objective advantages. A publicity campaign for a new fragrance, on the other hand, would be 98 percent emotional appeal. People choose a fragrance because it enhances their self-image and makes them feel more desirable, not because they get more cc's per bottle or because it lasts five times longer than other leading brands.

Next to each situation described below in the sentences, indicate whether the most important element of persuasion would be credibility (C), facts (F), logic (L), or emotion (E):

_____ 1. You are a supervisor concerned about the chronic lateness of your staff.

_____ 2. You are preparing a financial report indicating that a proposed expansion will bankrupt the company.

_____ 3. You are a senior maintenance person, showing a new employee how to resurface ice in an arena.

_____ 4. You are a student representative speaking to a college meeting on strategies for student recruitment.

_____ 5. You are trying to borrow your brother's car.

_____ 6. You are participating in a panel on drug abuse before a high-school audience.

_____ 7. You are negotiating a bank loan.

_____ 8. You are selling a honeymoon package to an engaged couple.

2 THE AIDA SEQUENCE

The elements of persuasion, in the combination you have decided is appropriate, help you determine the content of your persuasive message. The AIDA sequence—Attention, Interest, Desire, and Action—is a method of organizing your persuasive material in the most effective sequence. What follows is a general discussion of the four parts of the AIDA sequence. At the end of the chapter, you will find suggestions on how to create attention, interest, desire, and action in specific kinds of persuasive messages, such as sales messages, supervisory messages, complaints, and special requests.

Attention

The best-written, most eloquent message in the world will never achieve its purpose if no one is reading or listening. The persuasive message usually initiates communication, so it must work particularly hard to make the audience perceive some benefit in reading or listening. Of course, you want your audience to pay attention throughout your message, but **attention** in the AIDA sequence refers primarily to attracting and focusing the audience's attention at the beginning of the message. Consequently, attention-getting devices must be very brief—a few words or an image that the audience can take in before they make a conscious decision to start reading or listening. A range of possible techniques for attracting attention will be discussed in the sections on specific kinds of persuasive messages later in the chapter.

Interest

Once your audience has begun to read or listen to your message, you must reinforce their decision to pay attention by creating **interest** in the content of your message. The most effective means of creating interest is to discuss the topic people find most interesting—themselves. Your readers or listeners must find something directly **relevant** to themselves and their needs. The writer or speaker, on the other hand, is often too eager to get into the "real message" and start presenting information about his or her product or idea. Resist this temptation. The interest section should always contain the word "you" or "your"—preferably, more than once. Details about the decision you are trying to promote should come only after your audience have found a place for themselves in your message.

Desire

The **desire** section prepares the audience for a decision by moving them from self-interest to interest in the product, idea, or behaviour you are putting forward. The desire section is usually the largest because it contains the most information and greatest detail. It helps the audience translate interest into a definite commitment by answering all the important questions.

Action

If everything has gone as planned, your audience is completely convinced and is only waiting for directions to put this conviction into action. The **action** section tells the audience how to do this and, if possible, enables the audience to do it easily. Since the test of effective persuasion is some change of behaviour or belief on the part of the audience, the action section is the "make or break" part of the message. But without the appropriate buildup, the action section will lack an essential ingredient—the audience's decision to carry it out.

3 SALES MESSAGES

Attention

The decision to read a flyer, an advertisement, or a "junk mail" letter is made in a matter of seconds. The **attention** section must catch the reader's eye immediately. A quick look through a magazine illustrates several ways of doing this.

1. **An attention-getting picture.** A picture can attract your audience with something beautiful, like a landscape or a handsome face. It can be amusing, startling, puzzling, or provocative. The size of the picture can increase the impact, and so can the intensity of its colour. You can also use black and white pictures to create a contrast with competing images.

2. **A distinctive typeface.** You can also create visual impact by making the attention line stand out from the rest of the text. "Headline" type, coloured type, and other distinctive lettering can attract the reader's eye to the beginning of your message. This is particularly important if the reader might have difficulty reading the type used in the body of the message; for example, a large, bold headline might draw a passerby close enough to read a poster. Many people have to put on their glasses to read ordinary print, so an attention-getting headline will motivate them to make the effort to see the rest of the message.

3. **"Magic" words.** Free. Bargain. New. Opportunity. Sex. These are some of the words that most people find irresistible. Beginning a message with one of them almost guarantees at least momentary attention.

4. **A question.** Have you had your soup today? The question mark acts as a psychological prompt. The reader feels he or she must respond. This device is even more effective when the reader is unsure of the answer: "Do you make these common driving errors?" Who could resist finding out what they are?

5. **A provocative statement.** One out of every ten people reading this sentence will die from cigarette smoking. Like the question, the provocative statement presses a psychological button. It appeals to strong motives or emotions such as greed (You may already have won), fear (Your best friends won't tell you), or pity (Maria has never seen a BMW) and usually adds an element of curiosity.

6. **Humour.** Some of our best engineering ideas are gathering dust (vacuum cleaner ad). Often a humorous headline reinforces a visual image.
7. **Group identification.** Denture Wearers!—If you don't wear dentures, of course you won't read on. But if I'm selling denture adhesive, I don't care. My target group will be attracted. Use headings such as "Parents," "College Students," "Brides-to-be," or "Homeowners" to alert your target audience that your message is relevant to them.

The attention line needs room to work. It should not compete with other elements of equal size. Don't assume that if one attention-getting technique is good, three or four will be even more effective—more likely, they will cancel each other out. A glance through a local paper will usually give you examples of amateurish advertisements overloaded with attention-getting devices. Attention is broken up and dissipated instead of being focused at the beginning so that the rest of the message can do its job.

The following is a list of effective attention lines from magazine and newspaper advertisements. Identify the technique used in each one: "magic" words (MW), a question (Q), a provocative statement (PS), humour (H), or group identification (GI):

____ 1. Are you sensitive to moods?

____ 2. If television is a medium, then ours is well done.

____ 3. What every woman should know about the oil business.

____ 4. We're proud to offer you less than the competition.

____ 5. The difference between a shattered windshield and a shattered life.

____ 6. Sale of the century!

____ 7. Are you part of the breakfast success story?

____ 8. Every day, thousands of women perform surgery on their feet.

____ 9. If you're losing your hair …

Interest

Once you have your audience's attention, you have the opportunity to stimulate their interest in your message by making it directly relevant to them.

A sales message usually offers a product or service to the reader. Making sales depends on creating a need for that product or service. Since most people will not immediately perceive their need for, say, chimney relining, silk flower arrangements, or tickets to a benefit concert for unwed fathers, your job is to relate what you are selling to a need your audience is aware of.

The vast topic of human need has been rather neatly summarized by Abraham Maslow in a triangular diagram, as shown in Figure 7.1, with which you may be familiar.

We need not concern ourselves with the base or apex of the triangle, but a glance through any magazine will show you that virtually every product or service advertised is tied in to our need for safety and security, belonging, or esteem. **The interest section focuses on one of these needs, and encourages the reader to identify with this need.** This section focuses primarily on the reader, not what you are trying to sell. It will lead the reader from his or her natural self-interest to an interest in what you are promoting.

In order to write a good **interest** section, you need to spend some time analyzing your product, service, or idea, and deciding what need or needs it could fulfill. Sometimes the answer seems very clear-cut; a smoke detector, for example, meets safety needs, while a gold pen or a designer scarf appeals to esteem needs.

FIGURE 7.1   Abraham Maslow's Hierarchy of Human Needs

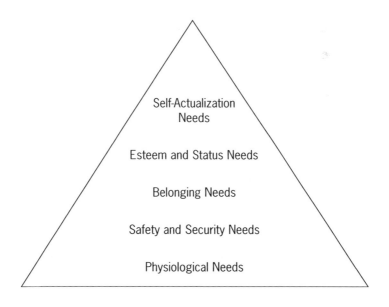

Other products give you a choice. A fitness club membership could meet **safety needs**—health, fitness, greater life-expectancy. It could also meet **belonging needs** such as getting on the fitness bandwagon, meeting new people, and looking more attractive. Finally, it could promise to put you ahead of the group rather than make you just part of the group. **Esteem needs** could be met by elite facilities, state-of-the-art equipment, a prestige location, or the promise of above-average achievement in appearance or athletic ability.

The important thing is to limit your appeal to your reader's needs to one or at most two categories of need. Remember that **belonging** and **esteem** needs are not really compatible; you buy "belonging" products because millions of Canadians use them every day, whereas you buy "esteem" products because they are the choice of the select few. A well-focused appeal to one need will be more effective than an attempt to be all things to all people. Analyze your target group as carefully as you analyze your product or service, to determine which need will be the most appropriate.

Help your reader identify with the need you are appealing to by emphasizing the words "you" and "your."

Examples

✗ POOR
Moneycheque provides a full payroll service to solve every payroll problem.

✓ IMPROVED
Your payroll problems are over when you use Moneycheque payroll service.

✗ POOR
Children will love our creative playground facilities.

✓ IMPROVED
Your child will love our creative playground facilities.

Desire

By the end of the **interest** section, your reader should have a clear idea of the benefit he or she can obtain from reading your message. Now that your reader is motivated to go on, you want to introduce the product or service you are trying to sell. The details you provide in the **desire** section should help your reader reach a decision to buy, or at least take definite action to this end. Examples of the kind of information that could be included in the desire section include price information, payment options, samples, pictures, technical specifications, or details of colour, size, material, and ingredients. The amount of information will depend on several factors. Ask yourself these questions:

1. **How big an investment is this for my reader?** Generally, the more something costs, the more the potential customers want to know before risking their money. If your readers are being asked to pay a premium price, they will want to know exactly what features make your product or service worth the difference.

2. **How complex is my product?** There is more to say about a microwave oven than about an electric kettle. When a product is both very expensive and very complex, like a car or a computer, it is unlikely that a single sales message can provide enough information to persuade a reader to buy. The purpose of the message will be to persuade the reader to find out more about the product or service.

3. **How significant are individual details to a decision to buy?** If you are selling stereo components or sports equipment, the technical specifications will be very important to your reader. A couple interested in a family vacation wants to know if children are welcome at your resort, and at what rates. Services such as child care, financial counselling, and education also require exact details to persuade readers that your organization is the one they are looking for. With some products, a competitive price is very important to a decision to buy, so the price must be included in the message.

On the other hand, an advertisement for mascara, premium Scotch whisky, or a rock CD gives very few concrete details because buying decisions are usually made on a subjective, non-quantifiable basis.

Action

If your desire section is successful, your reader is now ready to **take action.** This action could take several forms. The reader could actually purchase the product or service on the spot by phone or online, or take immediate action to write, phone, or e-mail for more information. Or he or she could decide to come to your store or office to buy or find out more about what you are selling. To write a good action section, you must first determine what action you want your reader to take. Then you must enable your reader to take this action as easily as possible. Finally, you must provide some motivation for doing so right away. Here are some techniques that will help you do this:

Making Action Easy

Phone/Internet Order or Inquiry
- phone number(s)
- toll-free number
- 24-hour line
- e-mail address/website URL
- credit cards accepted
- online payment

Mail Order
- address
- coupon, order form
- postage paid envelope/postcard
- credit cards accepted

Personal Sales
- address, map
- parking/transit information
- hours of opening
- credit cards accepted
- payment plan

Motivating Action
- free offer
- gift with purchase
- free trial
- money-back guarantee
- limited quantity
- limited-time offer
- deadline

The following examples, in Figures 7.2 and 7.3, show how **attention, interest, desire,** and **action** sections in a message can be combined for a unified persuasive effect:

FIGURE 7.2 Capital Inn Advertisement

76% of Business Travellers Say Their Expense Reports Are Being Scrutinized Like Never Before.

Times are tough. The economy is uncertain.
And your company's counting on you to keep expenses down.

Nobody understands the pressures of business travel like Capital Inn.

Capital Inns don't just look good on an expense report. You can depend on us for:
- an attractive room   • necessary business tools
- hospitality and comfort you need to relax and get the job done.

All at a very affordable rate.

For reservations at any Capital Inn, call 1-800-CAPITAL or your travel agent.

CAPITAL INN

FIGURE 7.3 MicroFresh Advertisement

MicroFresh Makes Vacuuming a *Mite* More Effective

Dust mites. No home is safe from these microscopic monsters. Most vacuums just blow the little critters from room to room. That's why Royal make the MicroFresh Filter System.

MicroFresh's electrostatically charged fibres capture particles as small as 0.1 micron—pollen, dust, even dust mites—leaving you vacuum exhaust 99.9% fresh.

So find out about equipping your vacuum with a MicroFresh filter system, and bag the bugs.

For more information, call *Royal Appliance* at 1-800-555-0821.

4 SPECIAL REQUESTS

In Chapter 4, you learned how to make routine requests, such as asking for product information or placing an order. Sometimes, however, you will want to make a request that does not appear to have any benefit for your reader. For example, you may want a guest speaker for a class, you may need information for a report or project, or you may wish to make special arrangements for production, payment, or delivery of some product or service.

Here is a special request sent by some Fire Protection Technology students to a well-known authority on environmental hazards. The students wanted a speaker on this topic for their urban geography class. They were also planning to tape the presentation as part of the research they were doing for a fire-protection project. The students were surprised and disappointed when Ms. de Silva turned them down, even though they had contacted her well in advance. Where did they go wrong?

Dear Ms. de Silva

The FPT students at Beaver College are looking for a speaker for their PYG 181 class on March 8, 2008. Are you available? The topic is cities. It's at 2:30 in room 1433. We would be very appreciative if you could make it, because our first-choice speaker backed out and we're supposed to tape this for a project. Thank you for taking the time to read this.

To make this special request effective, the students should have followed the AIDA sequence, with special emphasis on the **interest** section. The following letter uses the AIDA sequence appropriately to make a persuasive special request.

> Dear Dr. Hackenbush
>
> I am writing on behalf of the International Business students at Rouge River College to request your participation in our Near East Symposium.
>
> We are familiar with your recent book, <u>Dialling for Dinars: Telecommunications Opportunities in Kuwait</u>, and we would value the opportunity to share your knowledge of this area.
>
> The symposium will be held on Monday, January 25, from 3 to 5 PM. There will be three other speakers beside yourself, discussing the topic "Can Canada Create Lasting Markets in the Near East?" I have attached an information sheet on the other participants. The symposium will take place in the Lester Pearson Studio, and we anticipate an audience of about 150 from the college community and the general public.
>
> If you could let us know your decision before January 10, this would give us time to prepare the program.

Begin your message by focusing **attention.** Usually, you are writing to a specific audience, so you can be reasonably certain your message will be read. A clear opening will orient your reader to the purpose of your message and lead into the reader benefit of the **interest** section.

The **interest** section here, as in the sales message, appeals to a specific need. In the letter to Dr. Hackenbush, the students appealed to her **esteem** needs. In a special request, the **interest** section really does the selling, since the audience will not really receive anything concrete.

The **desire** section presents specific details of the request. It contributes to the reader's motivation by presenting the necessary steps in a clear, complete, positive way. The **action close** provides clear guidance to the reader. Since you may need a reply, even if it is negative, so that you can make plans or find an alternative, you may wish to take the initiative in the action close.

Example

I will call your office on Monday, April 3, to confirm your decision.

Figure 7.4 shows a complete special request, written using the four AIDA elements.

FIGURE 7.4 Sample Special Request

SPARKWOOD COMMUNITY OUTREACH ASSOCIATION
FAX MESSAGE

TO: Mr. Sagar Mahabir, Manager   DATE: October 10, 2009
           Allied Property Corp.

FAX: (403) 555-1212   NO. OF PAGES
                                                  (including this sheet): 2

FROM: Beverly Malcolm

For the past three years, your company has generously donated space in Central Mall for the Parkwood Community Outreach Association's annual fundraising gift wrapping service.

This year we are hoping that you will make possible a new service that would promote greater local initiative and community involvement. This project would provide shoppers with an opportunity to wrap their own gifts by making space, materials, and assistance available to them. This would be particularly helpful for the many seniors living in the neighbourhood of Central Mall. The attached plan shows how the location and layout of last year's gift wrap centre could be used for this alternative project. The dates and times would be similar to last year's gift wrapping service. If you have any other questions, I would be happy to discuss the project with you in detail.

Please let us know your decision before October 15 so that we can organize our publicity campaign. I am available at the Parkwood Community Centre, Monday to Thursday 8:30–4:30, at 555-4837.

Beverly Malcolm

Beverly Malcolm, Director

5 MAKING A COMPLAINT

You may think you already know how to complain. If you think of a complaint as primarily a means of letting out frustration, you probably have all the skills you need. But if you see complaining as **problem-solving**, a way of correcting an unsatisfactory situation, then you will need to use the persuasive strategies that we have been discussing in this chapter. Once again, the AIDA sequence can structure your message.

CASE 1

"Dear Chez Celeste," the letter began, "I've had some lousy restaurant meals in my time, but the one I experienced with four business associates in your restaurant last week was definitely a contender for All-Time Worst." Nicky felt a sort of stabbing pain starting behind her eyes. Maybe she should have looked over the weekend deposit statements before she started on the morning mail. In fact, Nicky was able to discover a number of vital tasks that had to be attended to right away, before she could get back to this particular piece of correspondence.

Finally, late in the afternoon, she picked it up again. Where did she leave off? Oh yes. At "All-Time Worst." She went on with a sigh. "It's a toss-up, but I'd have to say the food was worse than the service. When I pay over two hundred dollars to entertain business associates at lunch, I expect more than tiny portions of undercooked swill. And I expect professional service. You can bet that it will be a frosty Friday in July before I ever show up at Chez Celeste again. Disgustedly yours, Lorne Briarly." Nicky tried to visualize a group of five businesspeople at lunch last week. Was it Tuesday, the day the computer was down and two waiters never showed up? Hard to say what the actual problem was that was bothering Mr. Briarly. Since he was never coming back, it really didn't matter much. Nicky tossed the letter on the "To Be Answered (sometime)" file and picked up some order forms from Pastry Time.

Lorne may have felt better after writing this letter. But if his purpose was to be compensated for his unsatisfactory meal or to improve the food and service at Chez Celeste, he failed to achieve it. The following guidelines will show you how to adapt the AIDA sequence to write an effective complaint that will achieve results:

1. **Focus *attention* by identifying yourself and the subject you wish to discuss.**

   Example

   Last Saturday, September 8, I telephoned your store to find out the date of your "Fall Festival of Value."

   Include any dates, names, and other information that will help your reader know exactly what you are referring to. In the case above, Nicky would have been more sympathetic to Lorne Briarly's problem if she had been able to tie it to a day when she knew there had been a number of problems at the restaurant.

2. **Do not put your reader on the defensive by stating your problem right away. Create *interest* by appealing to the reader's needs:**

   Example

   Several regular customers at your store had recommended the "Fall Festival" as an outstanding sale.

   This tells your reader that, by solving the problem, he or she can re-establish a good relationship. This approach is much more motivating than threats or abuse.

3. **The *desire* section outlines your problem. Be as specific and objective as possible. Your reader needs facts to make a decision.**

   Examples

   POOR
   When I got there, I discovered that your salesclerk had grossly misrepresented the sale. Nothing was reduced except some out-of-season junk in huge sizes.

   IMPROVED
   When I arrived on September 10, I found none of the raincoats, skirts, or sweaters that your salesclerk said would be on sale. The only reduced items were summer sportswear in larger sizes.

   By describing his or her **feelings,** the first writer makes it easy for the reader to decide that he or she is fussy, unreasonable, and unrealistically picky. The second writer presents **evidence** that must be taken more seriously. Avoiding negative words and insults also keeps the reader from becoming angry and defensive. And of course, if you should discover that your complaint is not justified, a calm, logical tone will save you a lot of embarrassment.

4. **Tell the reader what *action* you expect.** This is a major distinction between complaining as therapy and complaining as problem-solving. The **action** section ties into the **interest** section by implying that your audience can restore your original feelings of goodwill with the appropriate action. Goodwill provides an incentive for your audience to take the action you suggest. Even if your suggestion is not accepted, your reader has an opening position for negotiation.

Example

I would appreciate your personal assurance that your next "Festival of Value" will live up to its name.

You will find another example of an effective complaint in Figure 7.5. Note how the writer has explained the problem in a clear, objective way and has given the reader an opportunity to correct the problem without losing self-respect.

6 SUPERVISORY COMMUNICATION

Unless you are planning a career in the armed forces, you will soon find that it is not possible to get things done simply by ordering people to do them. Of course, every workplace has rules and procedures, but an effective supervisor does not spend his or her day enforcing them. Instead, a good supervisor uses the techniques of persuasion to encourage workers to decide for themselves that doing the job in the prescribed way is the best way to achieve their own goals and the goals of the organization.

Attention

In a memo or e-mail message, the subject line is the attention line. Techniques like provocative statements, group identification, and "magic words" will work as effectively here as in a sales message.

Examples

Subject: Promotion Opportunities in Accounts Receivable
Subject: Free Trips to Europe
Subject: Improving Office Security

FIGURE 7.5 Sample Complaint

COMMUNITY LEGAL SERVICES
32 Gordon Blvd. Nepean ON K0K 3R3

March 6, 2007

Sylvie Dubois, Manager
Doughnut World
139 Gordon Blvd.
Nepean ON K0K 3R6

Dear Sylvie Dubois

For the last three months, our office has had a standing order for a "Gourmet Doughnut Tray" every Friday afternoon.

Up until now we have been very pleased with the service and selection from Doughnut World. Last Friday's delivery, however, contained only three varieties instead of the usual six or eight. In addition, the doughnuts were not as soft and fresh-tasting as we have come to expect from your store.

I have enclosed the statement for our February billing. I would appreciate a credit for one week's doughnuts on our next invoice.

Sincerely

Joelle Williams
Administrative Assistant

Of course, the subject line must also clearly identify the subject of the message. Questions, jokes, or other verbal gimmicks will not do this effectively. Because the subject line describes the content of the message, it should also help the reader see the relevance of the message.

The computer offers many eye-catching fonts and images that could, theoretically, be used to attract the reader's attention. These are not yet widely accepted in workplace communication. Stick to visual devices such as boldface, capital letters, and underlining, and do not use colour, pictures of popping champagne corks, or more than one font.

Interest

As we have seen, a good subject line makes the message relevant to the reader. The interest section should build on this by targeting a need, just as a sales message does. The interest section should be you-centred, using the words "you" and "your" and presenting the situation from the reader's point of view.

Example

✗ POOR
The photocopying budget is already 35 percent over projection. The department would save a lot of money on photocopying and repairs if large jobs were sent to the printing department.

✓ IMPROVED
You can avoid lineups and downtime for photocopying repairs by using the photocopier for small jobs of 20 copies or fewer. Larger jobs should be sent to the printing department, which will copy and collate your documents for you at far less cost to the department.

If you are targeting safety and security needs, use positive incentives wherever you can. State rewards for the behaviour you want to encourage, rather than negative consequences of the behaviour you want your reader to avoid.

Example

✗ POOR
If you do not pick up your new sticker by November 1, you may lose your parking space.

✓ IMPROVED
Picking up your new sticker by November 1 will guarantee your parking space.

Sometimes, of course, the negative consequences would be so serious that it is appropriate to warn your audience more explicitly. Encourage your reader to make a responsible decision by emphasizing "you" and "your."

Examples

✗ POOR
Operating the equipment without protective goggles can cause eye injury or even blindness.

✓ IMPROVED
Protect your eyes from injury or possible blindness by wearing protective goggles whenever you operate the equipment.

✗ POOR
Anyone who forgets to turn off the heater before leaving will be fined a day's pay.

✓ IMPROVED
Don't risk losing a day's pay by forgetting to turn off the heater before you leave.

Desire

The desire section turns from the reader to the decision you want him or her to make. It is usually the longest and the most detailed section.

Action

It is important to spell out the action you want your reader to take at the end of the message. Don't assume that it's obvious.

Example

✗ POOR
Details on the workshops available are posted on the department website. Please give this your attention.

✓ IMPROVED
Details on the workshops available are posted on the department website. If you would like to attend a workshop, please leave me a message before April 11.

FIGURE 7.6 Persuasive Memo

> Subj: Workplace safety alert
> Date: Jun /13 /2009 4:18:04 PM Pacific Daylight Time
> To: All employees
> From: Parisa Kabiri parisa.kabiri@netmail.ca
>
> You may have noticed Agriculture Canada inspectors in the office last week. They alerted us to the presence of Pharoah ants in our building. Besides being a nuisance, these ants can spread dangerous bacteria.
>
> On the weekend a pest control company will be spreading toxic bait in the basement. They expect that it will take approximately six weeks to make the building ant-free. In the meantime it is very important to keep all food out of the office, to ensure that the ants find nothing to eat except the bait. Otherwise we may need repeat treatments.
>
> Please do not eat or store any food in your office until you receive a message that all the ants are gone.
>
> You can store lunches or snacks in the staff fridge as usual but please eat outside the building. More information about the ants and the treatment being used can be found at www.dwfpest.com. Thank you for helping to keep our workplace clean and safe.

7 PERSUASION IN ORAL COMMUNICATION

Persuading face-to-face or over the telephone follows the same basic AIDA format.

Attention and Interest

Because you are establishing direct contact with your audience, you don't have to work as hard to attract **attention** as you do in a written message. The opening of your message should lead quickly into an area of interest to your audience. Remember that an introduction such as "Mrs. Blotz, I'm calling you about Slamtite Aluminum Screen Doors," will probably elicit the response "So what," even if your audience is too polite to say it out loud. Here are some better opening strategies, geared to specific needs.

Examples

Now that the warm weather is here, you may be concerned about letting more fresh air into your house without letting in flies and mosquitoes.

Are you concerned about the appearance of your front lawn?

Last winter you signed up for our Japanese flower arranging course.

Asking a question, or pausing for a word of agreement from your audience, helps to reinforce your listener's involvement and interest. It can also help you avoid wasting time if your audience indicates that they are not part of your target group; for example, if you begin by saying, "Is your house prepared for the heating season?" and your listener replies, "No, but who cares? I'll be in Florida all winter," there is not much point in pursuing the advantages of weather stripping.

Desire

The **desire** part of your message will again consist of specific details about your product or service. Remember that the average listener can absorb far fewer details than the average reader can. Limit yourself to the main features and advantages. Encourage your listener to ask and answer questions. Exploit the advantages of personal contact and participation that spoken communication offers. Don't turn your message into a monologue.

Action

Once your audience is satisfied with the amount of information they have received, suggest positive **action.** This may mean placing an order, making an appointment, or taking some other appropriate step. Let the listener know what action you expect. Offer specific alternatives; for example, if the customer says, "I can't put it on my credit card; my line of credit is all used up," you can offer to reserve an item or a space until the customer's cheque arrives. If your listener will be away for the Fall Fundraising Festival, perhaps he or she can volunteer now for the Spring Bazaar. An assertive (not aggressive) attitude at the **action** stage is an important element in establishing your credibility and, by extension, that of whatever it is you're selling.

Imagine that you are approaching Leisure World Health Club to ask them to donate a door prize to a student social event. Leisure World is located near your campus and offers fitness, sports, and spa facilities. Use the Planning Worksheet on page 169 to plan your persuasive telephone call.

PERSUASIVE TELEPHONE CALL–PLANNING WORKSHEET

Attention: Here is an opening line that will (1) ensure that I am speaking to the appropriate person, and (2) identify the purpose of my call in an interesting way:

_____

_____

_____

_____

Interest: What need am I appealing to?
How can the action I am proposing help my audience meet that need?

_____

_____

_____

_____

Desire: What details will my audience need to make a decision?

_____

_____

_____

_____

Action: What specific action is required?

_____

_____

_____

_____

Complaining in Person or by Telephone

Written communication is the best medium for complaining about a product or service, unless urgency demands immediate action. Within an organization, however, complaining orally is usually preferable, until the "official warning" stage is reached. A personal exchange clears the air more quickly, without leaving a permanent record to stir up old resentments. Because oral communication is more personal, however, some guidelines, like the ones below, are important.

1. **Try not to complain while you're feeling angry.** "Get on it right away" only if the situation requires **immediate** correction. Otherwise, give yourself time to cool off.
2. **Plan your message to achieve a specific purpose.** Blowing off steam is not an appropriate purpose. Decide on the action you are seeking. Assemble the facts and logical arguments that will convince your audience to give you the compensation or make the change that you want.
3. **Use the AIDA sequence.** Identify yourself, if necessary, and the general subject of your concern. Try to find something positive to begin with. Create an atmosphere of respect: two adults speaking. Give the facts. Indicate the action you are seeking.
4. **Be prepared to explain more fully, but don't get trapped into repeating the same points over and over.** One of you will surely lose your temper. If you are dealing with a subordinate, you may have to move from a discussion to written documentation. End the discussion by offering to put your concerns in writing for your employee's consideration. If you are getting nowhere with a colleague or a representative of a company or organization, ask to speak to the next in command. Don't waste time and energy flogging a dead horse.
5. **If your request is granted, be gracious.** Thank your listener for his or her help and cooperation. End on a note of restored goodwill.

CHAPTER REVIEW/SELF-TEST

1. The four elements of persuasion are
   (a) attention, interest, desire, action.
   (b) credibility, facts, logic, emotion.
   (c) sales, special request, complaint, direction.
   (d) product, place, promotion, price.

2. The Attention section of a persuasive message
    (a) can be anywhere in the message.
    (b) must be something purely visual.
    (c) is sometimes omitted.
    (d) must do its work in less than a second.

3. The Interest section of a persuasive message
    (a) discusses a subject the audience is interested in, like politics or music.
    (b) is usually the longest part of the message.
    (c) creates awareness of a need the audience already has, like security.
    (d) creates a need for a product or service.

4. The Desire section of a persuasive message
    (a) shows how the action will help the audience meet a need.
    (b) is usually the longest part of the message.
    (c) gives details needed to make a decision to act.
    (d) all of the above.

5. The Action section of a persuasive message
    (a) helps the audience act on their decision.
    (b) adds more details to convince the audience to decide.
    (c) creates awareness of a need.
    (d) is usually the shortest part of the message.

6. Which of the following would not be an appropriate device for attracting attention to a sales message?
    (a) a picture of a cute kitten
    (b) the intended reader's name
    (c) a paragraph about a cute kitten
    (d) the words "Opportunity Knocks!" in red letters

7. Encourage your audience to act on a buying decision by
    (a) making it clear how to contact or visit you.
    (b) making it convenient to pay you.
    (c) offering an incentive for quick action.
    (d) doing all of the above.

8. The Interest section of a special request
    (a) is optional.
    (b) creates awareness of safety and security needs.
    (c) usually targets esteem needs.
    (d) tells the audience how interested you are in receiving his or her assistance.

9. An effective complaint
   (a) gets the problem off your mind so you can move on with your life.
   (b) punishes the person who caused the problem.
   (c) brings about an appropriate solution to a problem.
   (d) lets the audience know that he or she has made a dangerous enemy.

10. As a supervisor, you will find that your most effective persuasive technique will be
    (a) the AIDA formula.
    (b) negative reinforcement.
    (c) reminding people that you are the boss.
    (d) asking nicely.

Answers

1.B 2.D 3.C 4.D 5.A 6.C 7.D 8.C 9.C 10.A

EXERCISES

1. Select an example of persuasive writing, such as a direct-mail letter or a newspaper or magazine advertisement. In approximately 250–300 words, identify and evaluate the **attention, interest, desire,** and **action** sections of the message. Attach the letter or ad to your critique.

2. You are assistant director of Eddie Haskell Community Youth Centre. The centre is setting up an after-school program, which will run from 3:30 to 5:30, five days a week, to provide activities for elementary school children who might otherwise be going home to an empty house. Prepare a promotional flyer describing your program, to be sent to parents of children at local schools. Use the AIDA sequence.

3. Write a persuasive message to be handed out by you at January registration, inviting your fellow students to use your personal income tax return preparation service. Use the AIDA sequence.

4. Compose an e-mail message to be sent to each student graduating from your college this semester, inviting him or her to join the college's alumni association at a special rate for new graduates.

5. Your college is collecting used clothing for disaster relief (your choice of disaster). Prepare a persuasive message calling for contributions, to be posted on bulletin boards throughout the school.

6. As a result of the promotional effort in question 5, your college has collected 1.5 tonnes of used clothing. Write a special request to Air Canada, asking the company to donate their services to fly the clothing to an appropriate destination.

7. Your organization has taped a program of community poets and singers in performance. Write a letter to a local cable television station, requesting that your program be aired.

8. Recently you lost marks for handing in an assignment late. The problem arose because of technical problems with the internal e-mail system at your college. Compose a message to be sent to the Information Technology Department at your college.

9. You recently went out to dinner with four friends after a late-afternoon sports event at the college. You chose nearby Matt and Perry-O's. However, you were very disappointed by the food and service. Write a letter to the manager with your complaints. Request appropriate compensation.

10. Write a letter of complaint to a company or other institution about a product or practice that has caused you concern. Request appropriate action.

11. You are the manager of Acme Collections. Your employees spend the entire day at their desks, writing or phoning delinquent debtors, eating doughnuts, and drinking coffee. You have approached the manager of Chez Fitness, a health club in your building, to arrange a company membership. If at least 15 employees sign up, they will receive membership benefits at one-third off the regular price. Write a memo persuading your employees to sign up.

12. You are the office manager at Homeguard Security Services. You are concerned about the amount of computer time that is being taken up by employees who are booking travel, buying on eBay, or chatting online. High-speed computer access is important to your workplace, and you do not wish to have to spend time and money disabling the equipment. Compose a memo to your employees persuading them to limit their use of personal computer time.

13. Write a critique of the following telephone sales message. Identify any weaknesses in applying the AIDA sequence effectively:

    Salesperson: Hello, Mrs. Lay-o?

    Customer: It's pronounced Lao, as in "now."

    Salesperson: I'm calling from East End Chimney Relining to see if you'd like to have your chimney relined. We're offering a special package on flues, masonry, grout, and dampers this month.

	All our work is guaranteed, and if you have any problems or your fireplace doesn't work or anything, we'll come right back and fix the problem. We use the finest-quality materials, and our prices are no higher than those of other chimney services. Remember that a defective chimney can lead to death by fire or asphyxiation.
Customer:	Well, then, I'm glad I don't have a fireplace.
Salesperson:	Well, thanks again, Mrs. Lay-o, and if you change your mind we're in the Yellow Pages under Chimney Builders and Repairers.

14. Approximately 30 graduates who pledged to your college alumni fund have not made their donation. Using the planning worksheet on page 169, plan a persuasive telephone call that will encourage them to meet their pledges.

CHAPTER EIGHT

ORAL COMMUNICATION WITHIN THE ORGANIZATION

Chapter Objectives

Successful completion of this chapter will enable you to:

1. Describe the importance of oral communication on the job.

2. Deliver effective oral presentations.

3. Participate effectively in meetings and small groups.

4. Participate effectively in teleconferencing, videoconferencing, and Web conferencing.

1 INTRODUCTION: THE IMPORTANCE OF ORAL COMMUNICATION

Why do we need to study oral communication? Talking seems to come naturally, except perhaps for public speaking, something most people do not do very often. The truth is that oral communication is an integral part of most jobs. Personal interaction is the central focus for salespeople, health-care workers, educators, counsellors, and many managers, but wherever people work together or serve the public in some way, listening and talking are an essential part of the job.

Chapter 1 discussed some advantages and disadvantages of oral communication, the barriers to effective oral communication, and strategies for overcoming these barriers. This text has also discussed oral strategies in most of its chapters. In this chapter, we discuss specific guidelines for effective oral communication within the organization. Topics include oral presentations, working in groups, and participating in meetings, both the usual face-to-face meetings and newer forms like teleconferencing, videoconferencing, and Web conferencing.

2 DELIVERING EFFECTIVE ORAL PRESENTATIONS

Few people look forward to giving oral presentations. A number of surveys have shown that public speaking is one thing people fear most—ahead of death! However, whether or not you fear public speaking, learning to do it effectively is vital. Making an oral presentation allows you to highlight your expertise, demonstrate your ability to use oral skills to achieve corporate goals, and be an active participant in the corporate arena. In addition, oral communication skills are valued by employers (see The Conference Board of Canada's chart on "Employability Skills Profile" on page xiii).

Most people never have to give formal speeches or address large groups at work. Oral presentations usually occur in front of small groups, like committees, or in formal or informal meetings involving only a few people. Here are some guidelines for delivering effective oral presentations in any setting.

Preparation

1. **Prepare, prepare, prepare.** Preparation may be the most important aspect of delivering a successful presentation. Obviously, you cannot always be completely prepared for unexpected questions in a meeting. However, being "more than ready" not only ensures that your knowledge of the topic is ade-

quate, but it also helps to calm your nerves. Your comfort with the material and your organized mode of delivery create confidence, which is very important for effective communication. It is usually painfully obvious to an audience when a presentation has not been carefully thought through and rehearsed, or when the presenter is uncomfortable with the material or presentation technology.

2. **Find out how long your presentation is expected to take.** Normally, a speech is delivered at about 150 words a minute. Make sure your material is adequate for the time allotted. This does not mean that a ten-minute oral report should be as dense as a 1500-word written report. You will need to minimize details and use repetition to ensure that you are getting your point across. However, rehashing points you have already made in order to fill up your allotted time is a sure-fire way to annoy and frustrate your listeners. Leave time for questions and feedback. If the audience doesn't respond, don't fill the gap by answering questions nobody asked. This suggests that you are having second thoughts about the organization and planning of your presentation. **Never take more than your allotted time.** It is always counterproductive to run overtime.

3. **Learn as much as you can about your audience.** Find out what they already know and what they need to know about your topic. Avoid telling them things they already know or presenting information that is inappropriate for their level of expertise (either too abstract or simplistic), especially if you are presenting technical information. If you are trying to persuade your listeners, some prior research into their values, needs, and wants is worthwhile, particularly if you are selling a product or idea.

4. **Decide what method of delivery is appropriate.** Very few situations call for reading or memorizing a fully scripted speech. In fact, both of these methods require a very talented speaker who can bring the script to life. In place of scripted speeches, presentation software programs such as PowerPoint or Corel are now commonly used. However, there are still times when presentation software is not suitable. The technology can be overused, especially if an audience has seen too much of it. Speaking with only a few point-form notes can be a much better choice if you wish to create an informal atmosphere or you wish to appeal more directly to your audience's emotions. If you are very confident and you know your material well, you can make a powerful impression by speaking without any notes. In all cases, audiences respond best if you "speak from the heart."

5. **Plan an overall structure.** For instance, a common pattern is to introduce your topic, make three to five main points, draw a conclusion, make recommendations, and then summarize all of these sections. To leave a lasting impression, some repetition is desirable. In your introduction and conclusion, always relate the material to your listeners' needs and interests.

Presentation Software

Presentation software has become the technology of choice in the workplace. Programs such as Microsoft PowerPoint or Corel give you several advantages:

- Slides provide a colourful visual focus for your audience.
- Your audience will both see and hear your most important information.
- Slides double as your speaker notes.
- You can easily incorporate graphics.
- You can easily make handouts.
- Presentations may be shared electronically.

Using the software is relatively easy. Without instruction, most users can quickly learn how to create basic slides. However, some of the features, such as graph and diagram creation, are best learned with some help from an expert.

Guidelines for Creating Presentations

1. Create a strong opening, possibly using a photograph, quotation, question, or other dramatic device. Organize your information carefully for simplicity, logical flow, and visibility. The most common error is to place too much information on a single slide.
2. Use contrasting colours so that text stands out against the background. Keep the text size at 22+ font size.
3. Reduce your main points to short phrases and sentences that provide a clear meaning for both you and the audience. The general rule of thumb is five phrases of about five words each. To avoid a repetitive, list-like quality, cluster two or three minor points under a subheading where possible. As long as the font size doesn't end up being too small, it is good to add pictures to illustrate your themes.
4. Plan to expand on each point (know your material well). Consider using preset animation, which allows you to bring in one point at a time. Both you and your listeners can easily get lost when all points on a slide are seen together, so it pays to bring in each new point just as you are ready for it. However, some types of information are best viewed all at once, for example, a graph and text that interprets the graph (see the example presentation, Figure 8.1).

FIGURE 8.1 Example Presentation Created with Microsoft PowerPoint

YOUR LIFE

Employee Lifestyle Study Results

By Robbie Pinkney
Organizational Research

D+M Manufacturing

Why did we ask about your lifestyle?

- A balanced healthy lifestyle
  - Reduces stress
  - Reduces illness
  - Helps family well-being
- Employers can help

> Our Employee Health
> 1995-2005 sick days rose by 20%
> Employee assistance plan stress counselling rose by 30%

How we did the research

- Held staff focus group
- Designed survey based on focus group
- Sent survey to all 978 employees
- 70% were returned
  - 558 plant staff
  - 87 office staff

Findings: Exercise Habits

- We need 30 minutes, moderate to vigorous, minimum 4 times per week
- Most of us do not:
  - Vigorous 8%
  - Moderate 19%
  - Mild 40%
  - None 33%

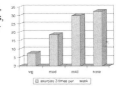

Findings: Eating Habits

- We need approximately 1200 mL (4 to 5 cups) of fruits and vegetables a day.
- Very few eat enough:
  - 5% eat 5 or more
  - 28% eat 3 to 4
  - 60% eat 1 to 2
  - 8% eat none

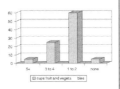

Findings: Family Time

- Family fun time (without TV or formal sport activity) is also very limited:
  - 4% had 6 hours or more
  - 32% had 3-5 hours
  - 40% had only 1-2 hours
  - 24% had none!

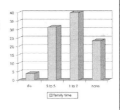

What does it all mean?

- A small number of employees are meeting their food and exercise needs.
- Family time is losing out for many.
- It is time to help!

Where do we go from here?

- Take two immediate actions:
  - evaluate cafeteria offerings
  - install a small gym
- Establish a lifestyle planning team
- Company will support by:
  - giving work time for meetings
  - helping to implement team's plans

5. Insert photos, clip art, graphs, and charts to enhance and replace text.
6. Use special effects, such as moving text, video, and sound, sparingly and judiciously. If you are using researched information to create graphs, or if you use graphs taken from websites such as Statistics Canada, give the source.
7. Prepare a strong ending. In most cases you will have a summary slide that pulls together the meaning of the entire presentation. In addition, you may choose to have a final slide that is designed to evoke discussion.
8. It is customary to provide handouts of your slides. Be sure to print off enough for everyone. You can save paper by printing more slides per page and making sheets two-sided. However, beware of reducing slides below a readable size. To decide when to distribute the handouts, think about whether your audience will want to take notes on them. Some people may need to remember details of your presentation. On the other hand, in some situations such as sales talks, you might prefer to discourage note-taking so that you can engage your audience with more eye contact. If you decide to hand out copies after the presentation, inform your audience that you will be doing so. Remember that your audience may also wish to have electronic copies of your presentation.

Guidelines for Delivering Presentations

1. Rehearse several times, with a friend if possible. Be sure of your timing. Arrive early and ensure the technology is working ahead of time. Be prepared with a backup in case of a technical problem, the ever-present disadvantage of this method. Don't make your audience wait while you open up the presentation. Have it ready to go.
2. Resist the temptation to look constantly at your screen, and beware of reading the points in a list-like manner. Reading is one of the most common errors. Never start with "OK" and end with "That's it." Be creative!
3. Take advantage of the technology's design, which allows you to face your audience at all times. To do so, look at your computer screen and not at the projection screen; and, of course, look up at your audience as much as possible. Do not make the common error of looking at the projection screen over your shoulder.
4. Always explain graphs and charts fully. Here you can approach the projection screen and use your hands, or use the on-screen pointers that can be created ahead of time.

5. If you have a discussion or question period, it may be important to leave your last slide in place. In addition, you may need to go back to some slides. Be prepared to do this quickly and graciously. Take care to exit from the presentation and bring closure.

Presentation Tips

1. **Use transition words to help your audience follow the presentation.** If you are making three main points, say so at the beginning of your presentation and then announce each point as you arrive at it. Let your listeners know when the end is coming. If they have drifted off or are watching the clock, announcing your conclusion brings them back to hear your summation.
2. **Use nonverbal communication to enhance your message.** Make sure your body language emphasizes what you are saying. Important nonverbal aspects of public speaking include the following:
   (a) **Posture.** Always stand up. Standing while others are sitting automatically confers authority and commands attention. It also helps you to project your voice to your listeners and to make eye contact with the whole group. Staying in your seat may make you feel secure, but this is a trap. A good presentation demands alertness and a bit of an edge. Good posture helps you breathe properly, which makes you feel and sound more relaxed. Do not pace or rock from foot to foot; your audience will find this distracting.
   (b) **Facial expression.** Your facial expression naturally provides a model for the audience's reaction to your presentation. You set the mood: audiences respond to your enjoyment and enthusiasm. Never appear apologetic for the nature or content of your report. Project involvement and conviction, but don't try to consciously arrange your face. Facial expression must come from within to be convincing.
   (c) **Eye contact.** Make brief contact with everyone in the group as you speak. Don't look at your computer screen any more than absolutely necessary; keep your eyes on the audience. Turn off equipment not in use; remove information from the screen that isn't relevant to the discussion.
3. **Use your voice effectively.** Several factors contribute to the impact of your speaking voice.
   (a) **Volume.** Speaking too loudly makes your message seem forced and impersonal, but this is not a common problem in public speaking. You are more likely to speak too softly or let your volume drop at the end of a sentence, especially if you are nervous. Eye contact with your audience will help you judge the volume that is both natural and clearly heard.

(b) **Pitch and emphasis.** When you are nervous, your voice rises in pitch. Some people have a habit of raising their pitch at the ends of sentences, projecting a sense of uncertainty. Make a conscious effort to lower your vocal pitch when you speak in public. Breathe deeply, and relax your shoulders whenever you pause.

(c) **Rate.** Nervousness, or even excitement and involvement, will cause you to speak more rapidly than usual. Some speakers deliberately speak quickly to convey the nonverbal message, "I know this isn't news to you/very interesting/very well thought through, so let's get it over with." Keep up an engaging pace, and slow down occasionally for emphasis or for clarity on difficult information.

4. **Project a positive attitude.** State your purpose without personal comment. Don't try to "lighten the atmosphere" by downgrading the importance or interest of your presentation. Others will accept your self-evaluation. Remember, too, that a casual attitude reflects negatively on the person or group that requested the presentation—often your supervisor or client.

5. **Adopt a "you-centred" attitude.** As pointed out in Chapter 2, think of your audience when you speak. Avoid using self-centred words like "I" or "we." If you are trying to persuade your audience, use the AIDA sequence discussed in Chapter 7.

6. **In a small-group setting, be prepared to move from the role of leader to group member once your presentation is over.** Some speakers find the transition from giving a presentation to participating in a meeting or discussion very difficult. Take your seat to indicate that you are leaving the "authority" role. If there is any negative feedback from the group, do not become defensive.

TIPS FOR RELAXED PRESENTING

Know that it is normal to feel nervous about presenting. The famous writer and speaker Mark Twain said, "There are two types of speakers: those that are nervous and those that are liars."

- Prepare your content thoroughly so that you are confident in your material.
- Try meditating beforehand: Close your eyes, breathe deeply, and visualize yourself speaking easily and happily. Wayne Dwyer, another famous writer and speaker, meditates before every presentation.
- Focus on your audience rather than yourself. Remember that your audience members are just people like you. Think about their needs. They will enjoy your presentation more if you are focused on their comfort and understanding rather than on being perfect yourself.
- Continue to breathe deeply and slowly before and throughout the presentation. If you are especially nervous, it would be worth your while to learn the technique of diaphragmatic breathing, which is taught in yoga classes and other venues.

Transparencies

If you do not have access to the equipment mentioned above, black-and-white overhead transparencies are easy to make. Create them on a word processor, making sure that the print is large enough for your audience to see, and print them. Place a blank transparency sheet in the paper feeder of a regular photocopier, and then proceed as if you were making a regular photocopy. With some machines, you can simply feed the blank transparency into a special slot so that making a monochrome transparency is even easier. To get colour transparencies, you may have to go to a professional copying firm, since many businesses do not have photocopying machines capable of producing colour copies.

Chapter 14 provides some examples of visuals that you can use to enhance your presentation. As well, Figure 8.1 provides an example of a simple and effective set of slides.

CONFESSIONS OF A SCARED SPEAKER

In my professional life, I speak publicly on a regular basis and people tell me that I seem confident and capable. Here is what they don't know. When I was a child, we had to give memorized speeches in school. I was so frightened that it seemed I became someone else at the front of the room, a person who could not see, hear, or think. I most certainly couldn't remember the words I had memorized so carefully!

Being a compliant student, I continued to give speeches as required over the years, and I always suffered extreme anxiety. When I had to speak once in front of 300 people, I found myself experiencing tunnel vision and nearly fainting on the way to the podium. Ironically, I was drawn to acting in high school and college plays. The glamour of the stage outweighed my stage fright. As a result, I had plenty of practice in overcoming my nerves.

Now, I have realized as an adult that other people are generally very anxious about public speaking. It seems most of us have a fear of looking "stupid" and I continue to experience this anxiety, although it has diminished to a large extent. I am so grateful for the brilliant invention of presentation software because it ensures that I remember my main points and it helps take the audience's eyes off me!

People will give you many suggestions about how to conquer your nerves, such as taking deep breaths or imagining your audience in their underwear. Some of these strategies will help you, but I can tell you that the only real way to gain confidence is to keep giving presentations. It is very helpful to take public speaking courses that give you weekly practice in front of other people. The instructor starts you with very short talks, and everyone supports you so that you gain self-assurance and skill. Eventually, you will feel more comfortable, so never give up!

3. MEETING AND WORKING IN SMALL GROUPS

Many jobs consist largely of attending meetings, sitting on committees, and working on group projects. Working efficiently with other people presents special challenges and requires well-developed social skills and an understanding of protocol for various situations.

Making Meetings Productive

Meetings involve productive exchanges of information and decision-making. They also are opportunities to demonstrate your professionalism and build working relationships. Since they are very expensive to companies, always consider whether a meeting is necessary. Could the goal be accomplished in another way, such as an e-mail exchange? If a meeting is needed, plan well to ensure it is productive. Routine business meetings should last no more than two hours.

Planning Meetings

Much of the responsibility for productive meetings rests with the chairperson. If you are chairing a meeting, your first task is to prepare an agenda. An example is shown in Figure 8.2. Think carefully about the most appropriate order for the topics to be discussed, because some discussions may logically follow others, and list each topic with an estimated amount of time to be spent on it. Consulting with others about the agenda helps to ensure that no important topic is forgotten; however, limiting the number of topics will keep the meeting focused and of reasonable length. Send the agenda to participants at least two days before the meeting, and include the minutes from the previous meeting, even if you distributed them earlier.

Choosing a Meeting Format

Meeting formats vary from informal chats to the very formal structure following *Robert's Rules of Order*. It is likely that you will choose a less formal procedure for meetings with colleagues. Whether formal or informal, a productive meeting is organized and leads to definite action. If you have attended formal meetings where Robert's Rules were in use, you will likely remember such terms as "seconding the motion" and "motion carried." Robert's formal rules are normally used when there are many people and there is a need for transparency. If you need to conduct meetings of this type, you can learn the rules by reading the book *Robert's Rules of Order.*

Conducting Meetings

As chairperson, you should start with a brief introduction, stating the goal and length of the meeting. Provide copies of the agenda for those who may not have one. If the group is not too large, make sure that everyone has been introduced. Since your job is to moderate the discussion, assign someone else to take minutes. A common practice is to appoint a time-keeper as well. Always start meetings on

FIGURE 8.2 Sample Meeting Agenda

**Georgian College
Student Activities Council**

Communications Subcommittee Meeting

Thursday, September 18, 2006
8 a.m. – 9:30 a.m.
Location: Room D304

**Invited: Subcommittee members, Fast-go Printing representative
Facilitator: Cathy Volpe ext. 1342**

Note: Please contact Cathy by September 12 if you cannot attend.

Introduction	Cathy	2 minutes
Minutes of the last meeting	Jarrett	3 minutes
Business arising from the minutes	Cathy	10 minutes
Updates on plan for outdoor clean-up day	Krishna, Tosha	20 minutes
Presentation of proposed student newspaper format	Fast-go rep	20 minutes
Discussion and vote on adoption of new student newspaper format	All	25 minutes
Other business		5 minutes

time. During discussions, your main tasks are to adhere to the agenda and ensure that everyone has a chance to speak—one person at a time. Never dominate the discussion or allow another person to do so. When a group decision is required, try to move the group toward consensus, once everyone has had an opportunity to express his or her views. It is your responsibility to bring the meeting to a close on time and arrange the way in which any unfinished business will be handled. Thank the group, and announce that the minutes will be sent out.

Participating in Meetings

If you are a participant in a meeting, you have a large role to play in making the meeting productive. Read the agenda ahead of time, finish any tasks assigned to you at the previous meeting, and gather relevant information wherever possible. During the meeting, listen carefully to other participants. When expressing your ideas, remain positive and courteous. Where there is disagreement, build on others' ideas rather than directly contradicting them. Always try to move the discussion forward and maintain a cooperative atmosphere. And remember, the chairperson requires your assistance in staying on topic and on time.

Recording and Sending Minutes

The term "minutes" may seem very old-fashioned, but the value of minutes never declines. A set of minutes is a record of what was said, what was decided, and what must be done after the meeting. Regardless of the procedural format you have chosen, a set of minutes should be taken at every meeting. Some groups require very detailed minutes while others want only decisions and action items to be recorded. Discuss with your group what is appropriate for your purposes. For easy recording and reading, it is best to use the same format at successive meetings. All minutes must include these items: the names of persons who attended or sent regrets, topics discussed, and actions to be taken with completion time lines, along with the names of the people who will take each action.

People will often have intentions of following through on tasks but forget in the rush of their work lives. Minutes sent out by e-mail soon after a meeting will remind everyone of tasks to be done. If there are no follow-up tasks to be done, minutes may be sent later with the next meeting's agenda so that people can save valuable meeting time by reviewing them beforehand. To lighten the burden of producing and sending out minutes, groups that meet regularly often rotate the responsibility. However, the meeting chair is ultimately responsible to ensure that the minutes are produced and distributed.

Working in Project Groups

Many occupations involve working together on projects. In developing a computer application, for example, a computer research lab might employ engineers, systems analysts, researchers, managers, support staff, and consultants. Students and visiting experts may round out the group.

FIGURE 8.3 Example of Meeting Minutes

Georgian College
Student Activities Council

Minutes

Communications Subcommittee Meeting
Thursday, September 18, 2006

Present: Krishna Dash, Dragos Ilas, Jarrett Smith, Cathy Volpe

Regrets: Tosha Vandenberg

Guest: Bill Nelson (Fast-go Printing representative)

Facilitator: Cathy Volpe, ext. 1342
Minutes recorded by Krishna Dash

Action	Member to take action	Completion date
Minutes from August 18 meeting were approved		
Council approved the plans for an outdoor clean-up day. A decision was made to write an article outlining the plan in the student newspaper.	Krishna will write an article and send to student newspaper editor	September 25
Council viewed the presentation by Bill Nelson, discussed the pros and cons of the new poster format and logo, and voted to adopt them. All voted in favour.	Dragos will follow up with Bill to arrange changeover of all print templates	Completion before next poster printing October 1

Next Communications Subcommittee Meeting:
    4 pm – 6 pm, October 17
    Student Council Meeting Room

Project groups are always in one of four stages:
1. **Inception:** A group accepts a project and sets its initial goal and strategies.
2. **Problem-Solving:** The group works out technical problems and procedures for attaining the goal.
3. **Conflict Resolution:** The group works through conflicts in points of view or interests and motives.
4. **Execution:** The group members carry out the technical tasks needed to reach the goal.

Stages 2 and 3 may be skipped if the tasks are very straightforward, but a group may have to return to earlier stages, perhaps repeatedly, if technical problems or conflicts arise.

A group must also pay attention to three main functions during all stages:
1. **Production:** A group needs to focus on the project goal, but not at the expense of the group well-being and member support.
2. **Group Well-being:** It is essential to make sure the group is working by establishing complementary roles and maintaining interaction.
3. **Member Support:** Each individual needs to feel included, valued, rewarded, and not unfairly burdened.

Here are some guidelines for getting along in a project group and making it productive:
1. Seek feedback from others regularly, listen to them carefully, and incorporate their suggestions whenever possible.
2. Always seek group approval before you proceed with a specific task on your own.
3. Always do what you said you would do, and do it on time.
4. Do not manipulate, dictate, show off, or slack off. Be an equal team member.
5. Avoid "group think," in which people feel they must always agree rather than propose alternative points of view.
6. When a problem arises, consider using the following problem-solving model with all group members:
    (a) Define and analyze the problem.
    (b) Brainstorm possible solutions, recording without criticism all that are proposed.
    (c) Agree on criteria for a solution.
    (d) Evaluate possible solutions using the agreed criteria.
    (e) Select a solution.

(f) Plan a course of action.
(g) When the action has been taken, evaluate its effect.

The following cases present typical problems that occur in meetings or project groups. Read each one, and suggest what has gone wrong. Compare your answers with those on page 193.

CASE 1

On Tuesday, Luda, director of programming at Newsoft Canada, received an e-mail from a client who wanted to speed up work on a new program. She called a meeting for the following day of programmers in the software development division. Since she was quite concerned about the fact that the project was not yet finished, she placed this item on the quickly drawn-up agenda: "Development delays." When she arrived at the meeting, she was met with coolness and averted eyes. She proceeded with the agenda in spite of her discomfort, but when it was time to discuss the delays no one spoke. Finally, the programmer responsible for liaison with clients broke the silence: "We will finish that program tomorrow. It took an extra week because the client called and asked for some changes about a month ago." What mistake did Luda make?

CASE 2

Robert works for a landscaping company. During a meeting about the design of lawns for a new public building, Robert said, "I think it might be important to create a windbreak with trees along the north boundary. I've gone into the building during construction, and the wind really rips across the main walkway." Ron said, "No, we haven't enough money in the budget for large trees." Robert felt a wave of resentment and said no more, but he wondered quietly why they couldn't plant young trees that would do the job after three or four years had passed. What mistake did Ron make?

CASE 3

Pierre was really excited about his proposal for a composting system outside the kitchen of the large restaurant where he worked as a sous-chef. He had done his homework and designed a system that would be inexpensive to build and easy to use. The finished compost would be used on the herb garden that was just outside the back entrance. He had been given a 20-minute slot in the monthly meeting to present his ideas and seek his coworkers' support in using the system. Pierre was so keen that he talked about the way composting works for 17 of his 20 minutes. He didn't notice the yawns, but he noticed the lack of enthusiasm the following week. What mistake did Pierre make?

CASE 4

Adina was invited to work with four other hospital staff on a project to brighten up the playroom in the pediatric wing. She was pleased because she felt she had a talent in interior decorating. At the first meeting, some ideas were brought up, but the group decided to think about possibilities for another week and meet again to make decisions. Adina couldn't keep her mind from picturing how it should be. She spent her two days off that week drawing up a plan for the room, including the colour of the paint, the type of furniture, and the selection of toys. At the next meeting, she handed out copies of her plan. The others glanced at the paper in their hands. After politely thanking her for her work, they carried on with discussing their ideas. She felt very angry that her efforts and talent were being ignored. What mistake did Adina make?

4 TECHNOLOGICAL COMMUNICATION

An organization may have offices, branches, staff, or clients in other parts of the country or the world. New technologies have brought about new meeting forms like teleconferencing, videoconferencing, and Web conferencing, which allow people to hold meetings without having to spend time and money on travelling. All the participants can now meet at the same time, wherever they are.

Teleconferencing

Teleconferencing enables more than two people to interact verbally via telephone lines. Meeting by teleconference has become commonplace because of the savings in travel time and costs. Like a face-to-face meeting, teleconferencing must be properly planned and conducted to be productive, and participants must be prepared to speak and listen effectively.

Planning and Conducting a Teleconference

The technical preparation for a teleconference varies depending on the system at your disposal. Always test the system before the actual meeting. Contact all the participants to make sure that everyone will be ready for the teleconference at the proposed time. Then, using the fax machine or e-mail, send participants copies of the agenda and other necessary documents so that they can be ready with useful questions and comments. During a teleconference with several participants, it is easy for them to get "lost" without visual contact, so you must work harder than in a face-to-face meeting to keep everyone involved.

Participating in a Teleconference

Effective performance in a teleconference demands listening skills that many people no longer possess. Most of us have learned to use the eye rather than the ear to take in new material. It requires a special effort to detect the feelings in a speaker's tone of voice and to concentrate on processing spoken information. When you contribute a comment, pay attention to the clarity and volume of your voice, and identify yourself so that everyone knows who is speaking.

Videoconferencing and Beyond

Videoconferencing allows a number of people to see and hear each other. Industrial projects can be so large that participants may live and work in different continents. Some companies are using this audio-visual technology to train geographically separate teams, such as sales teams, which earlier had to be brought together in one place — an expensive undertaking.

Videoconferencing

Videoconferencing equipment can cost more than $15 000 for a fully equipped conference room. If your company has chosen to purchase equipment, use it when face-to-face interaction is needed but travel costs are high. In organizing a videoconference, the same rules apply as with any other technology: know how it works, test it thoroughly, and prepare people by confirming their attendance and sending all the information they need ahead of time. Videoconferencing systems work in different ways, so the facilitator must be prepared to explain to participants what they must do, such as positioning themselves for the camera, pushing the right buttons before they speak, and using the mute feature when not speaking.

Web Conferencing

A variety of conferencing software has been invented for the purpose of group interaction. You may use these relatively inexpensive systems to write, talk, and work on the same documents in real time, all without leaving your computer. Some allow you to see each other as well. These systems are known by several names, such as groupware, Web conferencing software, and electronic meeting software. Recently, the line between technologies has been blurring as video and online conferencing technologies blend. However, a useful distinction is made by the terms "asynchronous" and "synchronous." Asynchronous technologies, sometimes called forums or bulletin boards, allow people to leave messages and documents that all members can access at a later time. Synchronous technologies, sometimes called chats or video-chats, allow real-time interaction. Each type has its own advantages and limitations and should be chosen with care.

Asynchronous interaction is generally good for in-depth discussions but poor for decision-making. Many work teams use a combination of both types, holding real-time meetings for decisions and asynchronous forums for ongoing communication and document sharing. Sometimes, these technologies are part of a company's intranet, a system that connects all employees electronically. It is important to make yourself aware of the capabilities of your organization's communication system.

Answers to Chapter Questions

Page 190:
1. Luda made the mistake of not consulting anyone in the department before setting the agenda. If she had, she would have discovered that there was no need for a meeting at all, and she could have avoided alienating her programmers.
2. Ron directly contradicted Robert rather than building on his idea. By doing so, he annoyed his fellow worker and missed a potentially good idea.
3. Pierre monopolized the time he had been given. When seeking the support of coworkers, it is essential to allow them time to ask questions and express their doubts or objections.
4. Adina proceeded with a task before seeking approval from the group. Some preparation for a meeting is desirable, but all members of a project group need to feel valued and included. If one person goes too far ahead on her own, she is bypassing the necessary group process as well as missing the ideas the group can generate.

CHAPTER REVIEW/SELF-TEST

1. Giving an oral presentation at work is a good way to
   (a) demonstrate your oral skills.
   (b) demonstrate your expertise.
   (c) control decision-making in a meeting.
   (d) a and b.
   (e) a, b, and c.

2. In order to ensure you have your listeners' attention
   (a) start your presentation by saying "OK."
   (b) use a dramatic device such as asking them a question or giving a startling fact.
   (c) take your time opening up the presentation so they will feel suspense.

3. Common problems with presentations are
   (a) too much text per slide.
   (b) not using preset animation to bring in one point at a time.
   (c) clustering of details.
   (d) a and b.
   (e) a, b, and c.

4. Special effects such as sound and moving text should be used
   (a) on most slides to entertain listeners.
   (b) on controversial slides to put listeners in a good mood so they won't challenge your ideas.
   (c) very sparingly to emphasize the most important aspects of your presentation.

5. Graphs and charts can
   (a) illustrate your ideas to make a stronger impact.
   (b) increase your listeners' understanding of difficult information.
   (c) confuse your audience if not fully explained.
   (d) all of the above.

6. When presenting orally, you should
   (a) face your audience and look at them over the computer screen.
   (b) keep your eyes on the computer screen so you won't be nervous.
   (c) read each point quickly and move on.
   (d) b and c.

7. Common voice problems are
   (a) trailing off in volume at the ends of sentences.
   (b) raising voice pitch at the ends of sentences, giving the impression of uncertainty.
   (c) speaking too quickly.
   (d) all of the above.

8. Meeting minutes should contain at least the following items:
   (a) names of persons who attended and who sent regrets.
   (b) topics discussed.
   (c) actions to be taken.
   (d) name of the person who will take each action.
   (e) all of the above.

9. Teleconferencing provides the chairperson with these advantages:
   (a) you can get other work done while attending the meeting and no one will notice.
   (b) you can save time and travel costs.
   (c) you can avoid preparing an agenda, consulting with other people on needed agenda items, or sending out minutes after the meeting.
   (d) a and b.

10. As chairperson of a videoconference, you must do the following:
    (a) know the technology well so that you can instruct participants in its use.
    (b) talk at least 50 percent of the time so that there is no lost air-time while people are thinking.
    (c) take time to ensure that everyone is positioned correctly for the camera.
    (d) a and c.
    (e) all of the above.

Answers

1.D 2.B 3.D 4.C 5.D 6.A 7.D 8.E 9.B 10.D

Exercises

1. In groups of three to five, choose a current decision that someone in the group is facing—for example, whether to complete the requirements for a diploma or take a job. Follow the problem-solving model on page 189 to help this person make a decision. You will need someone to write down the ideas generated during brainstorming. Every idea is recorded without judgment at this stage. Then proceed with the remaining steps.

2. Describe the characteristics of a team in which you have participated and that worked very well. Compare your list with that of a classmate, and present your combined list to the class.

3. Attend a meeting at your college or city council. Write a critique of the interaction you see. Submit it in memo format to your instructor.

4. This exercise will help you gain confidence. Give a one-minute speech describing your favourite music or musician. Follow the pattern of:
   (a) introduction, with an attention-grabber and your main point;
   (b) three characteristics of the music that you appreciate; and
   (c) a memorable closing about what the music means to you. Preparation may be done in class.

5. In pairs, select a product or service and develop a five-minute presentation to sell the product or service to your class. Keep their wants and needs foremost in your mind. Use the most up-to-date audio-visual equipment available to you to make a strong visual impact.

6. The following is a two-part assignment that gives you knowledge of Web conferencing systems, as well as skill development with Web research, evaluation, report writing, and meetings.

    (a) Work in pairs for this assignment. You both work for a company that has offices in Canada, Europe, and Asia. Your director has asked you to research Web conferencing systems that would allow international teams to function better at a lower cost. According to your career interests, decide what kinds of projects these teams would carry out, listing details of the tasks involved, such as planning meetings and working on technical drawings together. Find three different systems that offer a variety of features and compare them for their usefulness in accomplishing these tasks. Write a 700–800 word report evaluating the three systems and recommending one of them. (Refer to Chapters 12 and 13 for help with research and reporting.)

    (b) Continuing with your partner, prepare to hold a meeting to decide which of the above conferencing systems would be most appropriate for your company. Create an agenda and assign one person to be the meeting facilitator and the other to take minutes. In your next class, hold 20-minute meetings in groups of four. One pair will lead using their agenda and evaluation report, and then the other pair will lead. In the meetings, share the evaluation reports you have prepared and base your discussion on them. In the following class, submit two items to your instructor: your pair's evaluation report (working with it may show how it could be improved) and your meeting minutes. Your instructor may also ask you to evaluate each individual's participation in the meetings.

7. This exercise gives you practice in interacting positively with your colleagues and awareness of several errors often made by presenters. Change the following negative comments into constructive feedback, suggesting what to do in order to correct the problem. Eliminate all negative words such as "don't" and "couldn't." Your instructor may ask you to use constructive comments to evaluate other students' presentation.

    (a) Don't read your slides in a list-like manner.

    Example of constructive version: *Strive for varied pace and pitch. Talk about each point fully before moving to the next point on your slide.*

    (b) Don't start with "OK."

    Example of constructive version: *Plan a strong opening line to focus people's attention, avoiding expressions such as "OK."*

    (c) Don't talk through your nose. The tone is very unpleasant.

(d) Don't raise your pitch at the ends of your sentences as if you were asking a question. It makes you seem unsure of yourself.
(e) Don't rock from side to side.
(f) Don't wander aimlessly.
(g) Your slides are too crowded.
(h) We couldn't hear you from the back of the room.
(i) I couldn't read your chart. The text was too small.
(j) I couldn't follow your talk because what you were saying didn't match the slides I was seeing.

8. Develop an evaluation sheet for oral presentations that will be delivered in your class. List ten performance items based on the guidelines given in this chapter. Use the sheet to evaluate your own and your classmates' presentations.

PART THREE

THE JOB SEARCH

"Instead of my résumé, I've printed out my daily horoscope for the past year. You'll see that I'm a special person who's destined for great things!"

© 1998 Randy Glassbergen. www.glassbergen.com

CHAPTER NINE

RÉSUMÉS

Chapter Objectives

Successful completion of this chapter will enable you to:

1. Analyze your experience, aptitudes, and goals in preparation for writing a résumé.

2. Prepare an attractive résumé using a format appropriate to your experience and employment goals.

3. Prepare an electronic résumé suitable for scanning.

INTRODUCTION

Finding the right job will make a significant contribution to your happiness in life. The right job for you is one that uses your talents and expresses your values. It follows that in order to find the right job, you need to know what your talents and values are. You may have given them a lot of thought already, or you may have thought about them very little. This unit will help you analyze your experience to clarify your ideas about your future. Then you will want to match your goals to current job opportunities and, finally, present yourself to potential employers. This process is demanding, but the reward will be a job that offers satisfaction and achievement, rather than "the daily grind."

1 PRE-RÉSUMÉ ANALYSIS

The first step in finding a job is to spend some time thinking about yourself. This serves two purposes: first, it puts you in touch with your personal goals and helps you identify what you want out of life—what gives you satisfaction and a sense of achievement. Knowing these things will help you identify the kind of job you want and the kind of organization you would like to work for. Second, it helps you work out a strategy—in your résumé, application letters, and interviews—for presenting your skills and knowledge in an effective way to a prospective employer.

Several exercises may stimulate your thinking. You might like to try telling your "story" in an autobiography. In four or five pages, try describing your significant family experiences, activities you enjoyed and didn't enjoy, subjects you did well in at school and ones you had difficulty with, jobs you've had, and values and/or experiences that led you into your present course of study. When you are finished, look for patterns. Have you had consistent success or satisfaction with one kind of activity? Do you enjoy doing only the things you're good at? Are you conservative, or do you crave new experiences? Do you value external recognition, or is your own evaluation more important than that of other people?

Many psychologists who study life-span development believe that personal characteristics like these do not change significantly over time. You will probably be happier if you accept your strengths and limitations and look for jobs that are suited to them, rather than trying to change yourself in some radical way.

In addition to, or instead of, this narrative approach, you may like some more structured analytical activities. Here are some you might like to try:

Exercise 1

Activities Checklist

Circle any of the activities in the following list that you think you are good at. Add more activities at the bottom if anything important is missing. Then, using a different-coloured pen or pencil, circle the ones you enjoy doing.

fixing	thinking
teaching	designing
communicating	evaluating
writing	learning
analyzing	organizing
listening	motivating
coordinating	managing
cooperating	selling
counselling	decision-making
supervising	decorating
negotiating	leading
creating	performing
helping	planning
understanding	persuading
explaining	researching
reading	scheduling
observing	maintaining
problem-solving	speaking
coping	budgeting
investigating	building
directing	inventing

Exercise 2

From the previous list, identify the activities that you have circled as ones that you both enjoy and are good at. List them in the left-hand column of the following chart. In the right-hand column, write down situations that demonstrate you have the skills you listed.

Example

SKILL	EVIDENCE
leading	investment club president, successful minor hockey coach
scheduling	able to maintain B+ average while working part-time and participating in varsity sports

SKILL	EVIDENCE

Exercise 3

Think about activities that you enjoy. List five of your favourites in the left-hand column of the chart below. Opposite each one, try to identify what it is you enjoy about each activity.

Example

ACTIVITY	WHAT I ENJOY
Going to parties	Being with a large group of people, meeting new people, relaxed unstructured activity, eating and drinking, music

ACTIVITY	WHAT I ENJOY

Exercise 4

Personal Qualities

Rate each of the following qualities as they apply to you, using the following scale:

1. That's me
2. Describes me to some degree
3. Not me

_____ neat		_____ punctual
_____ accurate		_____ organized
_____ responsible		_____ outgoing
_____ cooperative		_____ helpful
_____ dependable		_____ tactful
_____ aggressive		_____ thorough
_____ conscientious		_____ efficient
_____ ambitious		_____ innovative
_____ hard-working		_____ decisive
_____ competitive		_____ self-disciplined
_____ enthusiastic		_____ adaptable
_____ imaginative		_____ assertive
_____ positive		_____ self-starting
_____ patient		_____ easygoing
_____ flexible		_____ quiet
_____ energetic		_____ relaxed
_____ supportive		_____ self-motivated
_____ sociable		_____ creative

Exercise 5

List the personal qualities that you ranked number 1 from the previous list. Write a paragraph justifying your choices.

Example

aggressive   assertive
ambitious   energetic
hard-working   competitive
decisive   self-starting

Whatever I do, I do to win. In sports, I practise by myself as well as with the team so that I can be the best on the field. I am working part-time and saving my money so that I can open my own business. I like to set concrete goals and work toward them. I enjoy my job as a commissioned salesperson because I like to know that how much I earn is directly related to how hard I work.

Exercise 6

Preferred Coworker Exercise

I prefer coworkers who are (check as many as apply)

_____ male   _____ in their thirties and forties

_____ female   _____ middle-aged

_____ both sexes   _____ a variety of ages

_____ in their twenties

_____ from a cultural background similar to mine

_____ from an educational background similar to mine

_____ from a variety of backgrounds

I prefer to work with people who

_____ like to work with objects or machines

_____ like to observe and investigate

_____ like to work with numbers or data

_____ like to use their imagination and creativity

_____ like to help or train people

_____ like to persuade or lead people

Exercise 7

Working Conditions Exercise

Use a blank sheet of paper to make a larger version of the table below:

Column 1	Column 2	Column 3
Jobs I have had	Things I disliked about job	Opposites
		Other positive things

In the first column, list all your jobs. In column 2 write down anything you disliked about the jobs in column 1, for example, low pay, unfair boss, or boring job. Don't worry about keeping these negative factors next to the right job. In column 3, write the opposite of the negative factor; for example, if you wrote "low pay" in column 2, put "high pay" in column 3. If you put "boring work" in column 2, put "interesting, challenging work" in column 3. When you finish, add any *positive* factors about your previous jobs that haven't appeared in the list in column 3. Then number your list of positive factors. Write the numbers down again on a separate piece of paper. Now go through the list, making a *forced choice* between each pair of factors. For example, if you listed

1. high pay
2. interesting, challenging work
3. regular hours
4. travel opportunities

you would decide which was more important to you in a job: high pay or interesting work. Then you would decide between high pay and regular hours, and then high pay and travel opportunities. Next compare "interesting, challenging work" with regular hours, and then with travel opportunities. Each time you choose one item over another, put a tick under that number on your separate piece of paper. For this list, you would have eleven tick marks distributed over four numbers. Count up the tick marks for each item, and rank the positive factors; for example, if "interesting, challenging work" received four ticks, "high pay" received three, and "regular hours" and "travel opportunities" two each, your "prioritized" list would look like this:

1. interesting, challenging work
2. high pay
3. regular hours ⎫
4. travel opportunities ⎭ tie

Exercise 8

Work Values Exercise

The following list describes a wide variety of rewards that people obtain from their jobs. Look at the definition of each satisfaction and rate the degree of importance that you would assign to it for yourself, using the scale below.

1. Very important
2. Somewhat important
3. Not very important

____ Helping Others: Involved in helping other people in a direct way, either individually or in small groups.

____ Public Contact: Frequent public contact with people.

____ Making Decisions: Power to decide the courses of action, policies, and so on.

____ Influencing People: Change attitudes or opinions or alter people's behaviour.

____ Working Alone: Do projects by myself.

____ Knowledge: Pursue knowledge, truth, and understanding.

____ Creativity: Create new ideas, programs, systems; not following a format previously adopted by others.

____ Change and Variety: Work responsibilities that frequently change in their content and setting.

____ Precision Work: Work in situations that require dexterity or attention to detail.

____ Stability: Job duties that are largely predictable and not likely to change over a long period of time.

____ Security: Assurance of keeping my job and a reasonable financial reward.

____ Excitement: Experiencing a high (or frequent) degree of stimulation in my work.

____ Recognition: Visible or public recognition for the quality of my work, so that people are aware of my accomplishments.

____ Profit, Gain: A strong likelihood of accumulating large amounts of money or possessions.

____ Independence: Being able to work without much intervention or direction from others.

____ Physical Challenge: Physical demands; speed, strength, stamina.

____ Time-Freedom: Work responsibilities that I can do according to my own schedule; no specific working hours required.

If you find this kind of analytic activity helpful, you will benefit from Richard Nelson Bolles's *What Color Is Your Parachute? A Practical Manual for Job-hunters and Career-changers* (Ten Speed Press, revised annually). This book has been described as "the bible of the job-search field."

2 PREPARING A RÉSUMÉ

Once you have increased your awareness of your areas of interest and competence, you will be ready to begin to put this information into a form that will be easily accessible to a potential employer. A **résumé** is a summary of relevant data about your qualifications and accomplishments. An appropriate format presents information in a concise, easy-to-read fashion. It shows that you are a person with excellent writing and presentation skills. A résumé must also reflect high ethical standards. Never "pad" your résumé by changing dates or job titles, exaggerating accomplishments, or adding nonexistent qualifications. Your résumé will stay in your file when you are hired. If any deliberate inaccuracies ever came to light they could be grounds for dismissal.

Choosing a Format

The **chronological** format is the résumé format you are probably most familiar with. The two examples in Figures 9.1 and 9.2 illustrate a typical chronological résumé. This format has separate sections for education, job experience, and other activities. Within the first two categories, the writer presents a year-by-year summary of his or her educational achievements and job responsibilities. Even if you decide not to use a chronological résumé for your job search, you might wish to prepare one for your own reference, as an organized record of your education and employment experience. The chronological format highlights **dates** and **job titles.** It is a good choice if you want to emphasize:

(a) the length of time you have spent in a particular job area,
(b) a consistent work history,
(c) a work history that shows progressive responsibility,
(d) your age.

For example, if you have a diploma in marketing and sales and have spent several years in a firm, starting as a salesperson and progressing to sales manager, and you now wish to move to a larger firm as a sales manager, a chronological résumé would be a good choice. It would be a poor choice if any of the following apply to you:

1. **You are looking for your first responsible job.** The emphasis on job titles will not be appropriate if the job titles are not obviously related to your career goal or if they are low-level titles like "salesclerk," "server," or "pizza delivery person." The description of duties performed will be pointless for many of these jobs; if this description is omitted, however, there won't be much to put in the résumé. Many important achievements—for example, as an athlete or student leader—will be relegated to the last part of the résumé.

2. **You are changing career goals.** If you are graduating in early childhood education but have learned that you hate kids and actually want to sell real estate, you don't want the title of your diploma to be the first thing a potential employer reads. Likewise, a consistent work history may not be an asset if you are trying to get into a new field. The reader's attention will be drawn to the job title, not the transferable skills.

3. **You have gaps in your academic or work history.** Because the chronological format emphasizes dates, gaps in the sequence will be very noticeable. If you made a few tries at your high-school or college diploma, have been unemployed, have worked in the home, or have changed jobs a lot, the chronological format is probably not for you.

If you fit any of these categories, a better choice would be the **functional,** or **skills,** format. This format, illustrated in Figures 9.3 and 9.4, emphasizes what you can do rather than when or where you learned how to do it. It integrates skills acquired through education, work, and other activities, so that achievements outside paid work get more recognition. Identifying your major skills leaves less work for your reader. Remember that during the initial screening a potential employer spends less than ten seconds on your résumé.

Preparing a Chronological Résumé

If you choose the chronological résumé, here are some pointers. The four headings correspond to the parts of the résumés illustrated in Figures 9.1 and 9.2.

Identification

Give your name, postal address, e-mail address, and phone number. If you have two addresses—for example, if you will be returning home at the end of the school year—include both with appropriate dates. The phone number should be a daytime number. Identify a pager or voice-mail number as such. If you are regularly out during the day and cannot receive calls or messages, add "after 6 PM" or whatever other information is relevant. Employers take a dim view of applicants who appear to be unreachable. Do **not** include your date of birth, sex, marital status, health, height, or similar data unless these factors are bona fide job requirements, which is very rare. Otherwise you are inviting employers to break the law by discriminating on the basis of age, sex, marital status, or disability. Do not put your social insurance number on your résumé. This number is confidential and should be used only on official personnel documents when you are hired.

Employment

Education or employment should be the next section; lead with the category that is likely to be the stronger factor in obtaining a job. List job titles and employers in reverse chronological order. Use years, not names of months. Put "part-time" or "summer employment," if applicable, beneath the job title, not in the column of dates.

Do not list job responsibilities if they are obvious—everyone knows a server takes food and beverage orders, brings food, takes payment, wipes off tables, and so forth. Note responsibilities only if the job title is vague (clerk) or if you performed additional duties (for example, if you were a salesclerk but acted as manager two days a week, regularly closing the store and making bank deposits). If you have held a number of similar jobs, do not list the responsibilities of each one. The repetition will make your résumé very long and very boring, without adding anything significant. Do not lock yourself into an elaborate structure that is not appropriate to someone at your career stage.

Education

List diplomas or degrees obtained in reverse chronological order; that is, start with the most recent. Put dates in years, omitting the names of months. Distinguish the name of the diploma obtained from the name of the institution

FIGURE 9.1 Sample Chronological Résumé

Allison Ghorbani
63 Pacific Wind Cres
Brampton ON L6R 2B1
(905) 555-1802 (cell phone)
aghorb@aol.com

EMPLOYMENT OBJECTIVE   A supervisory position in payroll management

SUMMARY   After completing an accounting diploma with a 3.5 GPA, I worked in a variety of business settings and discovered an aptitude for payroll management. In three years I earned two professional certificates and was promoted three times. Improvements I made to my current employer's payroll system have eliminated the need for costly seasonal outsourcing.

EXPERIENCE

2003–present   Chinguacousy Resource and Administration
Brampton ON
Senior Payroll Specialist (2005–present)
Payroll Specialist (2003–5)

2002–2003   NEBS
Cambridge ON
Payroll Clerk

2002   OfficeSolutions Temporary Placements
Cambridge ON
Clerical Worker

EDUCATION

2005   Canadian Payroll Association
Toronto ON
Payroll Supervisor Certificate

2003   Canadian Payroll Association
Toronto ON
Payroll Administrator Certificate

(continued)

FIGURE 9.1 Sample Chronological Résumé *(continued)*

1999–2002   Conestoga College
                    Cambridge ON
                    Business Administration–Accounting Diploma

ACTIVITIES AND INTERESTS

Posting indie music blog
Coaching girls' soccer team
Travel

REFERENCES

Supplied on request

---

by using underlining, shading, capitals, or some other visual device. Go back only as far as high-school graduation; omit even that, if it was more than ten years ago, unless you see it as a definite asset.

Do not list all the courses you took. The titles will mean little to your reader and take up a lot of space. Academic distinctions such as graduating with high honours or being on the Dean's Honour List can be noted on your résumé.

Other Activities

This is the place to list participation in sports, student government, volunteer organizations, and clubs. You can include interests and hobbies, if you wish. Remember that interviewers often use this information to get an interview started. Interests like "sports" or "music," shared by 90 percent of the population, don't give him or her much to go on. Use specific entries, such as "amateur weightlifting, coaching girls' soccer, cheering for the Blue Jays."

FIGURE 9.2 Sample Chronological Résumé

George H. Vandermeer
Apt 12 - 1233 Academy Dr
Windsor ON N9A 7G9
(519) 555-8938 (cell)
geevan@sympatico.ca

EMPLOYMENT OBJECTIVE
A management position in retail electronics or telecommunications

WORK EXPERIENCE

Sunset Radio Windsor ON	**Store Manager** (2004–present) **Management Trainee** (2003) **Sales Associate** (2002)	2002–present

Recommended for management training after two months with my current employer. Completed course with "outstanding" rating. Under my management, store has increased sales volume almost 35% and cut employee turnover in half.

All Geek to Me Toronto ON	**Service Technician**	2000–2002

Contributed to my employer's rating as "Best Computer Store" by *Now* magazine. Developed a significant customer base through referrals by satisfied clients.

The Phone Store Toronto ON	**Sales Associate (part time and summer employment)**	1999–2000

Time spent helping first-time cell phone buyers get the right phone and plan earned me two Employee Achievement Awards.

EDUCATION

Centennial College Toronto ON	**Computer Electronic Engineering Technician Diploma**	1999–2000

President's Honour List

ACTIVITIES AND INTERESTS
Keyboard player in band
Maintain web page for local youth volunteer agency
Enjoy sailing, camping, travel

REFERENCES
Available on request

Preparing the Functional Résumé

The headings in a **functional résumé** reflect the areas of skill you have to offer to a potential employer. These can be skills acquired in school, on the job, or through outside activities. A good pre-résumé analysis as described on page 201 is a prerequisite for a good functional résumé. Look over the exercises or other pre-résumé writing you have done. Think about the skills you have that would benefit a potential employer and that you would like to employ in your job. If you have a particular career in mind, try to identify the major skills required for success in that job. Some examples of skill areas are:

supervisory	creative
sales	management
computer	personnel
mechanical	problem-solving
clerical	accounting
child care	counselling

Depending on your education and career goal, you may choose fairly specific headings, like accounting or technical skills, or more general ones like flexibility, initiative, or people skills. Select the two or three headings that would be most important to a potential employer. The résumé in Figure 9.3 illustrates how to organize your education and experience under each heading. Practise creating your first entry in the space provided on page 215. Try to begin each line after the heading with a verb (for example, "successfully **completed** course in personal sales," "**supervised** four employees as head of work crew," "**organized** Student Awards Banquet for 200 guests as student council social convener"). Figure 9.4 presents another example of a functional résumé, which, though shorter than the one shown in Figure 9.3, is still effectively written.

If you will be applying for jobs in several fields, you may wish to change or reorganize parts of your functional résumé to include or emphasize different skills. Do not create a new résumé every time you apply for a job, however. A résumé that presents the skills you have to offer in an honest and appealing way should be appropriate for most jobs you are likely to be applying for.

Identification

The identification section of a functional résumé is the same as that of the chronological résumé.

Skill Heading

_____
_____
_____
_____
_____
_____
_____
_____
_____
_____

Education

Put in the year you completed your degree(s) and/or diploma(s). Give the name and subject area of the diploma or whatever you received, and the name of the institution. Do not include any other information because all relevant information should be somewhere under the skill headings.

Work Experience

List the years you worked at each job. Give the name and address of the company. Omit job titles, unless they are impressive. Do not list your duties because relevant information will be included under the appropriate skill headings.

Employment Objective

Two sections have not been discussed because they are optional features of either résumé format. The first is the **employment objective.** You may wish to prepare an employment objective statement **if you will accept only a specific type of job.**

FIGURE 9.3 Sample Functional Résumé

Madeleine Miranda
1053 Rushton Drive
Mississauga ON L5C 2E3
mmiranda@aareas.ca
(416) 555-9977

Employment Objective

A responsible position in industrial or retail promotions.

Skills and Abilities

Marketing and Sales Skills
— successfully completed courses in marketing, marketing research, and retailing
— received grade of "A with Distinction" in personal sales course
— "Employee of the Month" eight times in three years of part-time retail sales
— participated in marketing and advertising decisions
— responsible for merchandising and display in a variety of retail settings
— attended customer service programs offered to SportLine employees
— designed and implemented recruitment campaign that doubled membership of college racquetball club in two months

Management Skills
— successfully completed courses in management, organizational behaviour, and administrative communication
— prepared work schedules for eight employees
— trained and supervised employees
— president of college racquetball club
— vice-president of college student council, responsible for planning and implementing college-wide entertainment programs

Office and Computer Skills
— successfully completed courses in business communication
— type 55 wpm
— applied knowledge of Word and computerized inventory to school and work assignments

FIGURE 9.3 Sample Functional Résumé *(continued)*

Education

Sheridan College of Applied Arts and Technology   2007
Oakville, Ontario

Diploma: Marketing Administration

Work Experience

Pizza Buona Restaurant and Take-Out Inc.   2005–present
Whitby, Ontario

Night Manager

MaxiStores SportLine Division   2005
Weston, Ontario

SportLine Dufferin Mall   2003–2005
Toronto, Ontario

Night Manager/Salesperson (part-time)

Starr & Dean Sales   2003
North York, Ontario

Telemarketing Group Leader (summer employment)

Interests

Racquetball, field hockey, collecting Mickey Mouse memorabilia

FIGURE 9.4 Sample Functional Résumé

Sarah Yan Feng Chen
3432 Broadway W apt 4011
Vancouver BC V6R 2B3
(614) 555-2345
syfc1@aol.ca

OBJECTIVE
A career in the hospitality industry

QUALIFICATIONS
Diploma in General Business Studies
Successful work experience in a hotel
Ability to communicate effectively in writing, in person, and over the telephone
People orientation and commitment to service

EDUCATION
Diploma   Kwantlen Community College
            Surrey BC
            General Business Studies 2006

EXPERIENCE
Travel and Hospitality
- Completed one-semester field placement at the Coast Plaza Hotel, Vancouver
- Successfully applied knowledge of computerized reservation system, customer service, general office skills
- Travelled widely in North America and Asia

Communication
- Speak fluent English, Cantonese, and Mandarin
- Maintain personal web page
- Successfully completed courses in business communication and public speaking

People and Service
- Volunteer campus guide and tour leader for international students at Kwantlen College
- President, Surrey/Zuhai Sister City Friendship Committee
- Active in volunteer fundraising for United Way, Terry Fox Run, WaterCan, and other local causes

Examples

Employment Objective: A sales position with potential supervisory responsibility.

Employment Objective: A position in personnel or labour relations with opportunity for advancement.

If your mind is made up, this type of statement will save you from wasting time in interviews discussing jobs outside your field of interest. It is also useful if you are sending unsolicited applications to large companies with many employment areas (for example, a retail operation that has sales, marketing, accounting, personnel, and other departments). If, however, you are undecided about your career goals, an employment objective statement would be inappropriate. A vague, open-ended phrase like "a responsible position that uses my education and skills" is worse than useless, since it applies to virtually everyone in the world. Remember that you will want to have a résumé that can be sent out to many prospective employers. If your objective is too specific ("Mediterranean cruise director") you will get very little use out of your résumé. **Never** mention that your long-term objective is self-employment. You are telling the reader that you wish to learn the business from his or her company and then become part of the competition. This information will not get you an interview.

References

The other section you may wish to include is a list of **references.** Since references take up a lot of space, you may want to use just the phrase "References supplied on request." Have a typed sheet of references prepared to take to interviews. Always ask permission to use someone's name as a reference. If he or she does not agree immediately, find someone else. You want someone who will be unqualifiedly positive about you.

ESL TIP

If you plan to use a functional résumé, begin by creating a list of your educational qualifications and the jobs you have held. Add other significant experiences like volunteer positions or sports. Then make a list of three, four, or five skills that would be important to a potential employer in your field. Look at job postings online or in the newspaper if you need help identifying relevant skills. Look through your first list to find items that would show that you had these skills, and put them under the appropriate headings. Try to begin each item with a verb, in the present tense if you are doing something now, in a current job, or in the past tense if you did something in a previous job. Remember that communication skills are needed in every job. If you can communicate in more than one language be sure to mention this important asset.

Spelling or grammar errors on your résumé will give the impression that you do not know enough, or you do not care enough, to make it perfect. Check all proper names yourself, and then ask someone with excellent English skills to proofread what you have written.

Presenting the Résumé

The appearance and accuracy of your résumé are extremely important. During the initial screening of applications, an employer will spend approximately ten seconds on each résumé. What comes across in ten seconds?

1. **Paper.** This should be good quality, in a conservative shade. White is always a safe choice.
2. **Typeface.** Clear type in an easy-to-read style will be appreciated by your reader. Print from your disk on a letter-quality printer. Photocopy on a high-quality machine, or have copies made for you by a printer. Choose a conservative font. Italics and other fancy fonts are hard on the eyes.
3. **Layout.** A one-page résumé is ideal, but only if it fits easily on the page without overloading it. Lots of white space invites the reader into the page. A dense, crowded page discourages the potential reader. Leave generous left and right margins. Ensure that your spacing and margins are consistent. If you need two pages, try to make the amount on each page fairly equal. The layout is particularly significant to someone glancing at the résumé for an overall impression.

4. **Emphasis.** Quick reading will be easier if key information stands out. Add emphasis and variety by underlining, capitalization, shading, and other devices.

If your résumé passes this first test and is read more closely, other factors will determine whether you are called for an interview. Mechanical errors in spelling and grammar will almost always be fatal. Proofread rigorously, and then ask someone with good English skills to go over it again. Don't forget your name, address, and other proper names. Personnel officers report that this is where mistakes are most often made, because proofreaders assume these words are easy and therefore will be error-free. Your résumé will represent you to your potential employer. Everything about it should reflect your best effort.

3 ELECTRONIC RÉSUMÉS

Of course, your résumé does not have to exist only as a piece of paper. Putting your résumé online will enable you to apply immediately to a job posting. Some employers may ask you to submit your résumé on a form, somewhat like the older paper application form. You can see an example of an application form on page 240. In the future, it may be common to link your résumé to other documents that support your application, like a portfolio or newspaper article, or even to multimedia displays. If you have the expertise to create such a résumé now, it could put you on the cutting edge in fields where innovative thinking is valued. Do not attempt this unless you can create a high-quality, professional-looking presentation that accurately reflects what you have to offer to a potential employer. Avoid pictures of yourself. Employers do not want to be open to accusations of selecting applicants on the basis of race or physical appearance.

Whether your résumé is online or in print, an employer may choose to scan it electronically before deciding whether to give your application further consideration. Electronic scanning is frequently used by large organizations that receive many unsolicited résumés and that wish to develop a pool or database of candidates for openings as they arise. Any organization that receives large numbers of applications might choose electronic scanning as a labour-saving and objective way of identifying the most qualified applicants. Of course, a computer cannot "read" a résumé with the same judgment as a human being. All it can do is search for preprogrammed keywords, and identify résumés that contain a certain quota of them. Keywords are usually nouns, so unlike the typical résumé, particularly the functional résumé, a scannable résumé avoids phrases like "performed reception duties" or "organized activities for 40-member youth group" because these phrases get their punch from verbs—action words. To adapt your résumé for computer scanning, you can keep your existing skill headings and change the entries from verb phrases to nouns.

Example

CONVENTIONAL RÉSUMÉ
Communication Skills
— edited college newspaper
— created website for Student Council
— successfully completed course in workplace communication
— type 80 wpm
— speak fluent English and Cantonese

SCANNABLE RÉSUMÉ
Communication Skills
college newspaper editor
website development for Student Council
report writing course
typist: 80 wpm
English speaker; Cantonese speaker

You may add a "Keyword Summary" at the beginning of the résumé. This is a list of all the likely keywords from your résumé, separated by commas or periods. It should contain synonyms and other forms and tenses; for example, manager, management, managing, managed. Use both abbreviations and full words; for example, A/R, Accounts Receivable. Even if you correctly identify the important keywords in the job area you are applying for, it is impossible to know the exact form chosen by the human programmer.

Scanners get confused by fancy fonts or unusual punctuation such as bullets, slashes, or dashes. Since they read horizontally across the page, avoid a layout with parallel columns or strong vertical emphasis. Many features that work well in the typical résumé for human readers are not very effective for scanners. It is better to prepare a conventional résumé first and then create a separate scannable version if you think that a particular employer will be using this technology. If you are delivering a paper résumé that may be scanned, do not fold it. Use an envelope the same size as the paper.

CHAPTER REVIEW/SELF-TEST

1. Before you start writing your résumé you should
   (a) download a résumé template.
   (b) spend time thinking about your talents, values, and goals.
   (c) invest in a colour printer.
   (d) choose a standard format and stick to it.

2. The purpose of a résumé
   (a) is to convince an employer that you have the skills and qualities needed for success in the job you are applying for.
   (b) is to give a sample of your high standard of presentation.
   (c) is to provide a summary for your personnel file if you are hired.
   (d) includes all of the above.

3. The chronological résumé format is a good choice if
   (a) you have a consistent work history that is obviously relevant to the job(s) you are applying for.
   (b) you don't know what skills or experience an employer might be looking for.
   (c) you have only had one or two paying jobs.
   (d) all of the above.

4. The functional résumé is a good choice if
   (a) you wish to focus on what you have achieved in the past.
   (b) you are changing careers.
   (c) you wish to emphasize job titles and dates.
   (d) your most important qualifications come from paid employment.

5. The functional format
   (a) never includes dates or employment history.
   (b) always begins with a summary of qualifications.
   (c) must include skill headings clearly relevant to the job(s) you are applying for.
   (d) should include a list of your jobs and the duties you performed.

6. The Identification section of a résumé
   (a) should include the word "Résumé" in large letters.
   (b) should be no more than three lines long.
   (c) should have all the information the employer needs to get in touch with you immediately for a job interview.
   (d) is optional.

7. The Employment Objective section of a résumé
   (a) gives you an opportunity to describe your long-term career aspirations.
   (b) should be general enough to ensure that you will be considered for any available job.
   (c) should use the exact title of the job you are applying for.
   (d) limits the type of job you wish to be considered for.

8. The visual presentation of a résumé
   (a) should invite the audience to start reading.
   (b) is just as important as the contents.
   (c) must be flawless.
   (d) all of the above.

9. An e-mail résumé
   (a) should be as visually exciting as a paper résumé.
   (b) makes having a paper résumé unnecessary.
   (c) may include a keyword summary if you expect it to be scanned electronically.
   (d) all of the above.

10. Deliberate errors of fact on your résumé
    (a) are grounds for dismissal if they are discovered, even if you have been doing a good job.
    (b) are expected from people who are trying to show initiative and aggressiveness in looking for a job.
    (c) will not matter once you have proven yourself as an employee.
    (d) are not likely to be discovered.

Answers

1.B 2.D 3.A 4.B 5.C 6.C 7.D 8.D 9.C 10.A

EXERCISES

1. Select a partner in the class (try to find someone you do not know very well). Tell your partner about your top three achievements—things you have done that you are proud of. Ask your partner to write down the skills and personal qualities that each achievement demonstrates, in his or her opinion. When you have finished, change roles.

2. Select a partner in the class. Take turns interviewing one another about your personal histories. Here are some sample questions:

(a) Where were you born?
(b) How does coming from _____ benefit you?
(c) What is the most important value you learned from your family?
(d) Can you give an example of how this value applies to your future?
(e) What did you enjoy most about your experiences in school?
(f) What did you enjoy least?
(g) Why did you choose your present course of study?
(h) How has your college education changed your ideas about your future?
(i) Do you have a job? (*If yes:*) What is it?
(j) If the money was right, would you do this job full-time? Why or why not?

3. Prepare a list of keywords related to the kind of job you will be looking for when you have graduated. Include job titles, skills, education, computer applications, and any special terms related to your field.

4. If you have a paper résumé, go through it with a highlighter to identify the keywords currently there. Create a keyword summary using the words in your résumé and variations on them.

CHAPTER TEN

LOCATING AND APPLYING FOR JOBS

Chapter Objectives

Successful completion of this chapter will enable you to:

1. Research the job market and locate potential employers and job postings in newspapers, at placement centres, and through the World Wide Web.

2. Write an unsolicited letter of application.

3. Respond to a job posting effectively in writing.

4. Fill in application forms.

5. Respond to a job posting effectively by telephone.

INTRODUCTION

The process of preparing a résumé, if you have taken it seriously, will have given you a good insight into your goals in life and the skills and abilities you possess that will help you meet them. Now you have to consider what kind of employer needs those skills and can help you meet those goals. You need to identify what type of work you want to do and what kind of workplace you want to do it in. Finally, you have to identify specific employers who have this kind of job available. In the past, newspaper advertisements and personal contacts were the main source of information about job openings. New technology has made it possible to identify many more job possibilities. Regardless of how you learn about a job opening, you will need to approach the potential employer with an effective message asking him or her to review your résumé and give you the chance to be interviewed.

1 RESEARCHING THE JOB MARKET

Your résumé is now complete. The next question is, "Where do I send it?" Of course, you know about "help wanted" ads, but only about 20 percent of the jobs available are advertised in this way. If you are free to relocate, you can use the World Wide Web to search the help wanted ads in newspapers across Canada and in other countries. Other websites list jobs that may not have appeared in print. The Internet is a good recruitment tool for large organizations that often have many locations, experience high turnover, and are constantly hiring at the entry level.

You will also want to enhance your chances of finding the right job by tapping the 80 percent of jobs that are not advertised. This research can take two forms, depending on how firm your ideas are about what you want to do. You may have fairly fixed ideas by now, especially if you are graduating from a well-defined course such as cosmetics retailing, early childhood education, or marketing. If so, you will be interested in discovering potential employers. Ideally, this is a long-term project.

Newspapers, especially their business section, and community newspapers are a good source of information about new companies, expansions, new branch offices, and other changes that will open up jobs. Visit your college library regularly to look at professional journals and newsletters in your field, especially Canadian ones. These not only report new job opportunities but often give you an inside look at a potential workplace. Visit the websites of organizations in your field. If you do not know anyone personally who is employed in your field, consider making an appointment for an information-gathering interview as described in Exercise 2 at the end of this chapter. A company that has hired graduates from your program in the past would be a good place to look for someone to interview. Go to large events where you will have an opportunity to meet many people in your field. Visit local trade shows related to your future employment area. Watch for relevant conventions or professional meetings, especially ones that encourage students to attend by offering special rates.

Another way of obtaining an inside look is to talk to as many people as possible who are working in your field. The information gathered in either of these two ways will always be useful—if not in choosing an employer, then certainly in a job interview. An applicant who is knowledgeable and curious about the field will have a strong advantage.

When it is time to begin sending out unsolicited application letters, gather names and addresses from the Internet, trade directories (in your library), your college placement office, your personal contacts, and the Yellow Pages. Tell everyone—literally, everyone—you know or meet that you are looking for a job in a particular area, and follow up any leads. It has been estimated that 60 percent of all jobs are filled through personal contacts. If you have prepared yourself by keeping up-to-date on the job market in your career area, you will be able to be more selective about the employers' names you take from directories and personal contacts. You will also be able to write a more effective letter of application.

If you are graduating from a general program, such as business administration, or if you are contemplating a career change, you will want to research the job market in a broader way. You should be reading the newspaper and a variety of professional journals to learn not only about employers but also about occupations. Your college placement office should have reference books describing typical duties of various job titles. The National Occupation Classification, available on the Department of Human Resources and Skills Development website, describes over thirty thousand job titles organized by occupational groups, and is a valuable resource for understanding the Canadian labour market. A first-rate way of getting up-to-date information is to visit a company, social agency, or whatever interests you and talk to a supervisor who works there. This will not only give you an idea of the duties performed but also tell you about the personality factors that are important in that workplace. The exercises at the end of this chapter give more information about how to arrange an information-gathering interview.

Talk to employed friends and relatives about their jobs and employers. As you begin to focus on job areas that would be appropriate to your skills and values, you can start to use the research techniques suggested for job hunters who have defined their career goals. It is generally a mistake to choose a career simply because there are lots of jobs available in it, if it doesn't otherwise appeal to you very much. Job markets will change a lot in the 30 or 40 years you'll be working. You may already have observed radical changes in a brief time; one day employers are desperate for ticker-tape tossers, the next month ticker-tape tossers are being laid off in droves. If you are in a career because your personality and skills are suited to it, you will survive the ups and downs. If not, you won't; it's as simple as that.

ESL TIP

While an obvious form letter is not a good choice when you are applying for a job, it will be helpful if you create a simple template for your solicited and unsolicited letters of application, using the AIDA format. Then it will be easy to fill in the name of the job you are applying for or the area of employment you are seeking, and the skills, experience, or qualities that this employer is looking for. Do not copy letters from websites; some of them are good, but many are self-promoting in an exaggerated way. Create your own letter, and then have someone with excellent English skills help you proofread it. Because a covering letter typically refers to the past, the present, and the future, it will inevitably use many different tenses. Pay particular attention to the verbs in your letter when you are proofreading a final copy.

2 WRITING AN UNSOLICITED LETTER OF APPLICATION

An unsolicited letter of application is one that is not written in response to a job posting. It is a persuasive message, so it follows the AIDA sequence.

Attention

The best way to get **attention** is to use the name and title of the appropriate reader. Phone the organizations you wish to apply to, and get the name (with the right spelling) and title of the person who normally hires people for the type of position you are seeking. Focus your reader's attention by stating your employment objective in the first line of your letter.

Examples

POOR
Dear Sir:
I will be graduating from Amor De Cosmos Community College in June of this year and I have done a lot of research into employment opportunities in my chosen field.

IMPROVED
Mr. Gerald P. Slater
Pacific Micro Industries
Burnaby BC   V1C 2E3

Dear Mr. Slater
I will be graduating in Business Administration from Amor De Cosmos Community College this June, and I am writing to inquire about employment opportunities at Pacific Micro Industries.

If you have a personal contact in the organization who has suggested that you send an application, mention this in the opening of the letter.

Example
A fellow Lakeview College graduate, Kelly Baxter, now working as a resource quality technician in your division, suggested that I write to you to inquire about potential job openings at Metro Works.

Interest

Like any persuasive message, an application letter creates **interest** by appealing to a need. In this case, the need is not a personal one like belonging or esteem, but a corporate one. For example, a manufacturing company needs salespeople, a social agency needs people with interpersonal skills and experience, and a film production company needs people with technical skills. The more you have learned about a particular organization, the more specific your analysis of their needs can be and the more persuasively written your interest section can be. Naturally, you will want to emphasize reader benefit and reinforce it by using "you" and "your."

Examples

✗ POOR
I will be graduating with a diploma in fashion design and I am looking for an opportunity to gain experience in the fashion industry.

✓ IMPROVED
Your reputation as an industry leader is built on well-trained, dedicated staff; I believe my education and commitment would make me an asset to your team.

Desire

Create a **desire** in your audience to learn more about you by describing your particular qualifications. Do not repeat what is on your résumé. Draw your reader's attention to your résumé with a sentence like this one:

As you can see from the attached résumé, I have a diploma in retail floristry and three years' part-time experience as a floral assistant.

Use the rest of the paragraph to tell some things about yourself that do not find a place on a résumé, such as personal qualities. Do not be afraid to say that you are hard-working, flexible, outgoing, self-starting—if you are. It won't sound arrogant, just confident. If you give some evidence supporting your statement, as occurs in the following example, it will be even more effective.

During the past three years I have demonstrated my capacity for hard work and organization by maintaining a high grade-point average while holding down two part-time jobs.

Knowledge of the organization you are applying to should be demonstrated in a way that emphasizes reader benefit.

Since Chunky Chicken is expanding into Quebec, my language skills would be an asset to your firm.

If you have a particular interest in the organization or job you are applying for, it is a good idea to mention this, as long as you state it in a reader-centred way. After all, employers want motivated, enthusiastic employees.

Examples

POOR
Working at Rad would really be a thrill for me.

IMPROVED
As a lifelong Rad wearer, I would bring a sincere dedication to maintaining your image of quality and value to any job with your firm.

A well-written desire section should add a personal, more subjective dimension to the skills and achievements listed on your résumé. The flavour of your personality should come through.

Action

The purpose of a letter of application is to get you an interview. Always make a specific request at the end of the letter.

Example

I believe my education and skills could contribute to the success of Leisuretyme Health Spa. I would appreciate the opportunity to discuss this with you in person at your convenience.

If your schedule is not completely flexible, suggest a time, or a number of times, when you will be available. You may want to put your phone number here, even though it is on your résumé.

The letter in Figure 10.1 is an example of an effective unsolicited letter of application.

Several days after you have sent the letter, follow up by calling the person to whom you sent it. Identify yourself as the sender of the letter, ask if it has been read, and if so, whether you could come in for a discussion. This request serves several purposes. It will motivate your audience to read the letter, if they haven't already. If they are undecided about interviewing you, it may tip the balance, because it is harder to say no over the phone. If you are turned down, ask why. This may be painful, but you may learn something about the present situation of the organization and its hiring plans, about how your qualifications fit or don't fit the needs of that workplace, or even about the contents of your résumé and covering letter. Even if you don't get an interview, you will be gathering valuable information.

FIGURE 10.1 Sample Unsolicited Letter of Application

95 Borden Gate
North York ON M5Q 2G7
May 12, 2009

Mr. Jerome Doucette
Director, Sales Division
IBM Microelectronics Canada Ltd.
3900 St. Hilaire Ave. E.
Montreal QC H1C 2T9

Dear Mr. Doucette

I am writing to inquire about employment opportunities in the sales division of IBM Canada. As a graduate of the Business Computer Studies program at Seneca College and an IBM-PC owner, I can offer your company a thorough knowledge of IBM products and their business applications. In addition, I have completed courses in marketing and personal sales.

As you can see from the enclosed résumé, I have financed my college education by working part-time at Gentlemen's Tailors. While this experience gave me the chance to apply knowledge acquired through college courses, the most important thing I learned at Gentlemen's Tailors is I CAN SELL.

I believe that my education and sales ability would be an asset to your company, and I would appreciate an opportunity to discuss this personally with you in an interview. I am available any weekday before 2 PM.

Sincerely

James Gatti

James Gatti

3 ANSWERING A JOB ADVERTISEMENT

While you are researching potential employers through directories and personal contacts, you will, of course, be participating in the other job-seeking ritual: reading job postings in newspapers, college placement offices, Human Resource Centres of Canada, and on the Internet. While you will want to adapt good ideas from your unsolicited letters, the solicited letter of application has some different features. Begin by reading the advertisement or job posting several times, very carefully. A typical newspaper advertisement appears at the top of page 235.

Underline or highlight the words and phrases that would be particularly relevant if you were answering this ad. Remember that it costs money to run an advertisement. If an employer pays for space to include more information than the job title and the place to apply for it, it would be a fatal mistake to ignore this additional information. This may include both objective and subjective requirements. Objective requirements are measurable, for example, **previous sales experience, type 55 wpm, college diploma, CGA, speak Italian.** Subjective requirements are personal qualities such as **motivated, outgoing, self-starting, able to work without supervision, detail-oriented.** Pay particular attention to these subjective qualities; they give you valuable insight into how the employer perceives the job and the work environment. Compare the second and third advertisements on page 235. The objective requirements are the same, but the wording of the advertisements, particularly the personal qualities considered appropriate by each firm, reveals the different styles of the two workplaces. Perhaps you can see immediately that you would enjoy working at one place but not the other. In this case you need to write only one letter, showing that you are compatible with this job. If you feel that you could work in either environment, you could apply for both jobs, but you would need two very different letters of application.

Do not be tempted to falsify your personality. You may be flexible, but if you are a quiet, steady worker who prefers doing things on your own, with your own system, you would be very frustrated in a free-wheeling, team-oriented environment that required everyone to achieve consensus and then abide by the decisions of the group. On the other hand, if you like to socialize at work and be in a lively environment, you will soon grow depressed in a two-person office where you spend long periods of time staring at the screen. You may get a job by misrepresenting your personal qualities, but you won't keep it (or want to keep it) if you have to be someone you're not for 40 hours a week.

SHIPPING / RECEIVING / CUSTOMER SERVICE
$28 500

Nansen Laboratories, a leading manufacturer of veterinary drugs, requires a mature, self-motivated individual to fill the Assistant Manager position at our Regina location.

Experience incorporating shipping/receiving, customer service, computer entry is preferred.

Company benefits package available.

Résumé to Nansen Laboratories.

CASH AUDIT / ACCOUNTING CLERK

We are a ladieswear chain looking for a responsible individual who has an aptitude for figures and enjoys problem-solving. This person must be detail-oriented and capable of meeting deadlines. Familiarity with Microsoft Great Plains and Word required. Please send your résumé to ...

ACCOUNTING ASSISTANT

We are an expanding retail chain with an aggressive young management team. You possess the accounting, interpersonal, and organizational skills necessary to provide support for our busy operations. If you have Microsoft Great Plains and Word experience and are interested in career advancement, call ...

Assuming that the employee profile identified by the advertisement matches your attributes and qualifications and that you have marked in the ad the information you will need to write your letter, you can begin. The solicited letter of application is a persuasive message and follows the AIDA sequence.

Attention

Bring your letter to the **attention** of the appropriate reader by using his or her name and title. This may be stated in the advertisement. If not, or if the information is not complete, phone the company to get it. This is an important step; it demonstrates that you can take initiative and gives your letter a professional-looking opening. If it is absolutely impossible to get the name and title—for example, if you are replying to a box number or you can't afford to phone across the country to get a name—use a subject line instead of a salutation.

Example

Alberta Government Employment Office
10011–109 Street
Edmonton AB T5J 3S8

Subject: Application for position as Director of Market Standards, competition no. CCM942-3GM

Interest

By advertising a job opening, your potential employer has already identified his or her need. Begin by stating the position applied for, using the job title exactly as posted, and state where and when you heard of it. This will ensure that your application reaches the appropriate person. An organization may have advertised various job openings.

Example

I am applying for the position of Special Events Assistant Coordinator advertised in the April 21 Globe and Mail.

Remember to underline the title of a newspaper, magazine, or website.

Desire

Since you have prepared a résumé to summarize most of the exact details normally included in the **desire** section, make sure your reader takes a look at it. If specific qualifications are called for in the advertisement (for example, recreation graduate, previous experience working with the disabled), mention briefly that you meet these requirements.

Example

As you can see from the enclosed résumé, I have a diploma in Office Administration from Mountainview College. In addition, I have worked part-time for four years as a receptionist.

If you do **not** have a particular qualification, say nothing about it. Avoid negative sentences like, "Although I do not have any work experience in this field...."

If relevant experience for this particular job is not on your résumé, add a sentence describing it.

Begin a new paragraph to describe your personal qualities. If some are called for in the ad, make sure that you talk about them. Support your statements with evidence.

Example

As a sales assistant in a small sporting goods store, I was frequently left in sole charge. This required responsibility, self-direction, and lots of problem-solving skills.

If no qualities are specifically mentioned, you should still describe your strong personal attributes. Look at the section on creating desire in the unsolicited letter (see page 231) for some suggestions.

Action

Ask for an interview, following the suggestions on page 232 under **action** in the unsolicited letter. See Figure 10.2 for a reply to the first advertisement on page 235.

Faxing an Application

Often a job posting will ask you to fax your résumé. Should you send just the résumé, or should you include a letter of application? On the one hand, asking for a fax suggests that the employer is pressed for time and wishes to make a decision quickly. A lengthy letter may well be ignored. On the other hand, unless the job requirements are completely objective—which is usual only in low-level, mechanical jobs—your résumé alone cannot supply all the relevant information. You might like to compromise by sending a fax cover sheet with a brief message pointing out the relevance of your skills and experience to the job you are applying for. If you send a longer message, avoid laying it out like a letter to be sent through the mail. Give it a more open format like the message on page 160.

Applying for Jobs Online

You can use the Internet to apply for jobs in two ways. The first is to use it as a kind of electronic bulletin board by posting your résumé where you hope it will be seen by a prospective employer. Current statistics for this method of finding a

FIGURE 10.2 Sample Solicited Letter of Application

23 Applewood Rd.
Saskatoon SK S7B 1E9
May 10, 2009

Ms. Lorna Harris
Manager, Nansen Laboratories
2230 Millgate Pkwy.
Regina SK S4G 3K2

Dear Ms. Harris

I am applying for the position of Assistant Manager advertised in the May 9 Regina Leader Post. As you can see from the enclosed résumé, I will be graduating from the General Business program at Prairie College this June. Before returning to school, I had three years' experience as a shipper/receiver. College courses in data entry and computer applications, as well as customer service training as night manager at Cooperative Cable, have given me the experience you require.

As a returning student, I maintained a high B average while working at a demanding part-time job and doing volunteer work with the People's Food Bank. Your firm would benefit from my motivation, attention to detail, and excellent communication skills. I would welcome the opportunity to discuss this position with you further in an interview at your convenience.

Sincerely

George Kourakos

George Kourakos

job are not very encouraging. For example, workopolis.ca encourages employers to pay to search its database by pointing out that they have over one million résumés on file. The website jobhuntersbible.com estimates your chances of finding a job in this way as less than one-half of 1 percent.

A far better way to use the Internet is to look for job postings and apply to them electronically. Many company and government websites have a field where you can fill out or download an application form. Sites like monster.ca and workopolis.ca post job advertisements and allow you to reply to them by creating or attaching your résumé and forwarding it to the employer, with a covering letter if you wish. Remember that most employers posting jobs here will be screening for keywords.

4 APPLICATION FORMS

Application forms perform two functions for an employer. First, they assemble information about every employee's education, skills, and work history in a standard format. For this reason, you may be asked to fill out an application form **after** you have been offered a job, so that the employer can have this information officially on file. This application form will be part of your permanent record. If you have misrepresented any details on your application form, and this is discovered even years later, it could be grounds for firing you.

The second function of the application form is to create a streamlined résumé that can be quickly scanned by the person making the hiring decision. This saves time for someone looking for applicants with specific experience and education, usually for low-level jobs that do not require a wide variety of skills. Postings for these jobs often ask you to reply by telephone. If you pass the first screening, you will be asked to come in and fill out an application form. The employer may review your application on the spot and then interview you, or you may be asked to leave your application and wait to be contacted for an interview later. A hiring decision will be based mainly on whether your qualifications meet the requirements of the job, but of course, neat, accurate presentation will help your application make the best possible impression. Bring your chronological résumé with you to help you get the details complete and avoid having to cross out or add anything. Figure 10.3 shows you a typical application form filled out.

FIGURE 10.3 Sample Application Form

APPLICATION FOR EMPLOYMENT
(PLEASE PRINT)

Date: 08/06/06 (Day, Month, Year)

Name: Birdella Matthews
Address: 15 Muriel Ave. **City:** Toronto **Prov:** ON **Apt No.:**
Postal Code: M4K 3C3 **Telephone No.:** (416) 555-1212

PLEASE INDICATE DATES AVAILABLE: Immediately

EDUCATION

	Name of School	Course	Started Mth\Yr	Left Mth\Yr	Grade Completed
High School	Eastern Commerce		Sept. 03	June 04	12
Vocational School					
College\University	Seneca College	Gen. Business	Sept. 04	June 06	Diploma

Details of other courses completed: Computer Design, George Brown Continuing Education, 2005

Employment History: (Show most recent position first)

Employer & Address	Supervisor	Type of Work	Reason for Leaving	Date Started	Date Left
1. Black's Camera, Gerrard Square	T. Lam	Asst. Manager Retail	Currently Employed	Dec. 04	
2. Tim Horton's, 328 Pape Ave.	R. Richie	Food Service	Better Employment Opportunity	Sept. 03	Dec. 04
3. Pape Recreation Centre, 821 Pape Ave.	J. Ross	Recreation Leader	Summer Contract	June 03	Sept. 03
4. Acme Marketing, 231 Bathurst Ave.	E. Chau	Telemarketer	Return to School	June 02	Sept. 02

If applying for clerical\secretarial position:

List computer skills: Lotus ☐ W.P. ☐ Keyboarding Speed(wpm) ☐
Other (list)

If applying for warehouse position, indicate experience.

Receiving ☐ Order filling ☐ Packing ☐ Shipping ☐ Lift Truck ☐ Pallett Truck ☐

General Information:

Have you any outside business activities or part-time jobs? No   Describe

PERSON TO BE NOTIFIED IN CASE OF ACCIDENT OR EMERGENCY (Only if Hired)

NAME: Verna Matthews
RELATIONSHIP: Mother
HOME TELEPHONE NUMBER: (416) 555-1212
BUS TELEPHONE NUMBER: (416) 555-3800

It is understood and agreed that my employment is subject to receipt of satisfactory references. The Company is authorized to make inquiries concerning the information given hereon and I agree to release any person or organization from the consequences of answers to such inquiries. I understand that any deliberate misrepresentation by me on this application will be sufficient cause for dismissal should I be employed by the company.

SIGNATURE Birdella Matthews

5 ORAL COMMUNICATION: REPLYING TO A JOB POSTING BY TELEPHONE

Sometimes organizations prefer to screen applicants by phone. If so, the advertisement will invite you to call a number and ask for a particular person. The person taking the calls will probably ask you a few questions, describe the job, and then take your name or ask you to come in with your résumé for an interview, or to fill out an application. Since a phone conversation can't be revised like a letter if you make mistakes, and since phoning can be naturally stressful anyway, you will need to prepare yourself if you want to make a good impression. Here are some general hints:

1. Read the ad thoroughly, and look over your résumé just before you call.
2. Have the advertisement and your résumé in front of you for reference. Have pen and paper handy to note directions and write in your calendar or date book.
3. When you are connected, identify yourself and the purpose of your call.
4. Listen carefully. Ask questions if you are confused or need to know something important. The location of the job, whether it is full-time or part-time, and the type of business are important questions, but don't get into details about the dental plan.
5. Listen carefully to the questions you are asked. Answer as briefly as possible; this is a screening call, not an interview. Put yourself in a positive light; sound confident. Don't apologize or put yourself down, even as a joke.
6. If you are asked to come in, be prepared to suggest a time and date (that's why you need your calendar or date book handy). Get directions and write them down.
7. If you are not asked, but you think you might like working there, take the initiative and ask if you can bring or mail a copy of your résumé.
8. Thank your listener and hang up.

Here is a sample telephone application:

Reception:	Hello, Acme Associates.
Kris Chong:	Hello. May I please speak to Donna Ross?
Reception:	May I ask who's calling?
Kris Chong:	My name is Kris Chong. I'm calling about the advertisement for a Junior Programmer in this morning's *Toronto Star*.

Reception:	One moment, please.
Donna Ross:	Donna Ross speaking.
Kris Chong:	Hello. My name is Kris Chong, and I'm calling about the Junior Programmer position advertised in the *Star* this morning.
Donna Ross:	Yes, Mr. Chong. Could you tell me something about your educational background?
Kris Chong:	I'm graduating this month in computer programming and operating from Maple Leaf College.
Donna Ross:	Have you had any work experience with computers?
Kris Chong:	I've used a computerized inventory system at my part-time job at National Tire, and I demonstrated software at the computer show this spring.
Donna Ross:	Well, we're looking for someone who's familiar with Lotus and is prepared to do some shiftwork. We're located in an industrial plaza, so you would need your own transportation.
Kris Chong:	Would this shiftwork involve weekends?
Donna Ross:	Saturdays, but no Sundays. Would you be available for an interview later today or tomorrow?
Kris Chong:	I could come in after 11:00 tomorrow.
Donna Ross:	Fine, if you could come in at 11:15 tomorrow and ask for me. Please bring a copy of your résumé. I'll just put you back to reception and you can get the directions.
Kris Chong:	Thank you very much. I look forward to meeting you.
Donna Ross:	Yes, I'll see you tomorrow, then. Goodbye.

CHAPTER REVIEW/SELF-TEST

1. A majority of people find a job through
   (a) posting a résumé on the Internet.
   (b) personal contacts.
   (c) answering an ad in the newspaper.
   (d) applying online.

2. If you have already chosen your preferred employment area you should be
   (a) relying on job advertisements to find openings.
   (b) posting your résumé and waiting for employers to contact you.
   (c) reading and networking to make contacts and find openings in your field.
   (d) postponing your job search until graduation.

3. You can explore possible areas of employment by
   (a) informally interviewing people who are currently working in jobs that might interest you.
   (b) researching job descriptions on the Human Resources and Skills Development Canada website.
   (c) finding tests and career advice on the Internet.
   (d) doing all of the above.

4. Researching the job market
   (a) is important only if you do not know what kind of job you are interested in.
   (b) will help you appear more knowledgeable in an interview.
   (c) can be left until you need to apply for a job.
   (d) is not important when the economy is booming.

5. A covering letter accompanying your résumé
   (a) is a waste of your reader's time and should be omitted.
   (b) helps your reader see how your qualifications would meet his or her needs.
   (c) is only useful when you are applying for a job by mail.
   (d) explains why this job is important to your career plans.

6. An **unsolicited** letter of application
   (a) uses the AIDA format.
   (b) begins by explaining why you need a job.
   (c) is written in response to a job posting.
   (d) all of the above.

7. A letter applying for a job that has been advertised
   (a) should focus only on objective job requirements.
   (b) should repeat all relevant information from your résumé.
   (c) is called a solicited letter of application.
   (d) should point out any qualifications mentioned in the advertisement that you don't have, along with the ones you do.

8. A **solicited** letter of application begins
   (a) by explaining why you want this job.
   (b) by identifying the job you are applying for.
   (c) with an attention-getting gimmick.
   (d) with a brief overview of your qualifications.

9. Which of the following sentences would be appropriate in the Desire section of a letter of application?
   (a) "I have always been a loyal customer of Club Casino fashions and it would be great to get an employee discount."
   (b) "Although I have only worked part-time in an electronics store up to now, I think I can handle the responsibilities of a management trainee in retail fashion."
   (c) "Please look over my résumé and see if you have any jobs I'm qualified for."
   (d) "Your business would benefit from my retail experience and knowledge of your product line."

10. Which of the following sentences would be appropriate in the Action section of a letter of application?
    (a) "I would appreciate the opportunity to discuss this position with you personally in an interview at your convenience."
    (b) "Thank you for taking the time to read this letter."
    (c) "I hope I will be hearing from you soon as I am anxious to start working for your wonderful firm."
    (d) all of the above.

Answers

1.B 2.C 3.D 4.B 5.B 6.A 7.C 8.B 9.D 10.A

EXERCISES

1. Select a company or organization you might like to work for. Using the resources of your library, collect a nucleus of ten to fifteen newspaper or magazine articles about your company. Throughout the semester, check current periodicals to select at least five more articles. On the basis of your material, give a five-minute class presentation on the positive features of working there.

2. Select a company or organization you might like to work for. Phone the company and request an appointment for an information-gathering interview with an appropriate person. The purpose of the interview is to obtain information about employment opportunities for college graduates. Here are some sample questions:
   (a) Does your company hire community college graduates?
   (b) What kinds of jobs are currently held by college graduates?
   (c) What college programs prepare students for employment at this company?
   (d) What further training, if any, is offered?
   (e) What strengths do community college graduates bring to employment with this company?
   (f) Has this employer identified any consistent weaknesses in the preparation of college graduates?
   (g) What additional skills or education would enhance a college graduate's chances of promotion?
   Share the information you have gathered with your classmates in the form of a handout or informal oral presentation.

3. Using your clipping file and/or interview as a resource, write an unsolicited letter of application to a company or organization that interests you.

4. Take a page from the employment section of a newspaper that contains jobs for which you will be qualified after graduation. Make a list of all the personal qualities asked for. Identify those that are requested most frequently. Select five qualities that you think you possess. For each one, write a sentence beginning, "My _____ is demonstrated by _____."

5. Choose an employment advertisement from a newspaper, college placement service, or other source. Write a letter applying for the job, using your own qualifications.

6. Visit one of the following:
   (a) Human Resource Centre of Canada
   (b) college placement office

(c) employment agency

(d) company that offers a trainee program (e.g., for management or sales jobs)

In a five-minute oral presentation, tell your class what kinds of jobs were available and what the advantages and disadvantages are of using this method of finding a job.

7. From the newspaper files in your college library, select an employment advertisement for a job in your field that appeared about a month ago. Call or visit the company, and ask the following questions:

(a) Why did you choose a newspaper advertisement to fill this job?

(b) Did you use additional methods?

(c) How many applicants responded to the advertisement?

(d) Were most of them qualified for the job?

(e) Did the successful applicant learn about the job from the newspaper?

(f) Did the successful applicant have all the qualifications you asked for?

Report your findings to the class.

CHAPTER ELEVEN

INTERVIEWS

Chapter Objectives

Successful completion of this chapter will enable you to:

1. Prepare for an interview.

2. Demonstrate your knowledge and personal qualities in a job interview.

3. Follow-up on an interview.

4. Learn from an interview.

5. Perform your best on standardized tests.

INTRODUCTION

An employer has many ways of assessing the skills and qualifications of a potential employee. Besides supplying your résumé and letter of application, you may be asked to fill in an application form or write an exam or aptitude test. Sometimes employers request academic transcripts, letters of recommendation, or portfolios of your work. The interview, however, is a unique opportunity to learn about your personality and test your values and goals. A face-to-face meeting helps a potential employer see you as a whole person and evaluate your compatibility with the job and the people you would be working with. It follows that the best advice for an interviewee is "be yourself." But sometimes anxiety and stress can prevent you from being your best self, and you give the interviewer a false impression. Preparation is the best way to overcome nervousness and ensure that your interviewer gets a look at the real you.

1 PREPARING FOR THE INTERVIEW

CASE 1

Ian was starting to panic. Achieving perfect hair had taken a little longer than he'd planned, and then a broken shoelace added five minutes while he looked for another pair. In the end, he had to change shoes. Now he was down to his last litre of gas. He didn't want to risk his good suit at a self-serve, so he was forced to take an eight-block detour to a full-service gas station. Seriously behind schedule now, he finally found a parking spot several streets away from the head office of Richmond Textiles. Even at a dead run, he knew he'd be late for the interview.

Arriving, finally, out of breath and wind-blown, Ian managed to pant out his name to the receptionist. "Oh, go right in; they're waiting for you." This was it, then — his shot at the assistant manager's job he really wanted. Let's hope the interview went better than the rest of the morning.

Interviewer 1:	Hello, I'm Laura Murphy from Personnel. This is Amir Salloum, our operations manager.
Ian:	Hi! I'm Ian Chandler.
Interviewer 2:	Did you have some problems finding the office?
Ian:	Well, not really; I couldn't find a place to park.
Interviewer 2:	Oh, we have our own car park. If we'd known you were coming by car, we could have given your name to the attendant.

Interviewer 1:	Well, now that you're here, Ian, we'd like to get to know a bit more about you, and then tell you about the job we want to fill. It says on your résumé that you enjoy reading. What kinds of things do you like to read?
Ian:	Uh, well—I read the newspaper.
Interviewer 2:	There's certainly a lot going on these days. Are there any particular stories you're following?
Ian:	Well, I'm hoping Calgary makes the playoffs.
Interviewer 1:	Are you following any of the developments in the Middle East?
Ian:	No, um, not really, I haven't noticed much about that …

(half an hour later)

Interviewer 2:	Now, Ian, you must have some questions about the job.
Ian:	I guess I don't know much about textiles. I'm not much into fashion.
Interviewer 1:	Richmond Textiles makes industrial textiles. We're not in the fashion business.
Ian:	Oh. What exactly are industrial textiles? …

Ian had worked hard to complete his diploma in business administration with good marks. His résumé had been impressive enough to get him onto the short list of candidates to be interviewed for the job at Richmond Textiles. But poor preparation probably cost him the job when he got to the interview. How could Ian have improved his chances of success? Here are some guidelines Ian should have followed to prepare for a successful interview.

1. **Phone ahead to get directions to the interview.** Ask how long it should take to get there from your house by car or public transit, whichever you will be using, at the time you will be coming. If the answer seems hesitant or vague, check with your transit commission or someone who drives that route. Being late makes a terrible impression, so if this interview is important and the territory is unfamiliar, you may want to make a trial run. Arriving an hour early can look over-eager and will give you a lot of time to become nervous.
2. **Plan your wardrobe and check that everything will be ready to put on.** Don't discover a missing button or fallen hem on your way out the door. Choose a conservative outfit, not necessarily your "best clothes." A suit is always appropriate for men and women, so consider investing in one if you don't own one and expect to attend several interviews. Certain occupations—day-care

worker, entertainment industry accountant—may allow you to dress in a fairly casual way on the job, but "dressing up" for an interview in conventional business attire is a mark of respect and an indication that you take the interview and the interviewer seriously. Dress to daytime standards—minimal jewellery and cologne, low-key hair. For men, if you are willing to sacrifice your earring and goatee once you are working, you should do it for the interview, but if you intend to keep them you might as well give the interviewer the message right away. If the employer hates men with earrings, you wouldn't be very happy working there, anyway. For women, avoid up-to-the-minute high fashion unless you will be working in that field. Your outfit shouldn't attract attention away from you.

3. **Look over your résumé.** The interviewer will base most of his or her opening questions on it. It will create a poor impression if you seem unfamiliar with the information you have supplied. Never contradict or correct anything on your résumé. If your interviewer knows how to interview, however, you will not be asked simply to repeat information already there. Most interview questions try to get you to expand on the information—for example, "Why did you choose College X?" "How do you think participating in Varsity athletics helped you?" "What was your favourite job?" Look over your pre-résumé analysis to get you thinking about your values and goals in a more in-depth way.

Remember, one purpose of an interview is to get to know you better; this process cannot succeed if **you** don't know yourself very well.

If you have a "hobbies and interests" section in your résumé, give some thought to what you put there. Interviewers often pick this area as an icebreaker at the opening of the interview. If you mention sports, for example, you may be asked to comment on the prospects of a local team, or if you put travel or reading, a question about recent books or trips could be expected. Ian made a poor impression with his lame answers to questions about his reading. He should have predicted a question about what he liked to read and prepared an appropriate answer, such as "I like to follow the sports pages pretty closely. I'm also reading a biography of John Lennon." This would have kept him away from embarrassing questions about current events. If you do not have an "interests" section in your résumé, the interviewer may begin by asking you what you like to do in your spare time. If you are caught off guard, you may blurt out something like "Oh, knock off a couple of two-fours," which will waste an opportunity to look good. Prepare some answers about your leisure activities and interests in advance.

4. **Research the organization.** Another purpose of an interview is to give you information about the job and the organization. You will be better able to profit from this part of the interview if you have gathered some preliminary information. If the organization has a website, it should be your first stop. Follow up by searching elsewhere on the Internet, especially if you notice that the main website contains little information or has not been kept up-to-date. Check databases for newspaper stories and annual reports. Ask teachers and employed friends and relatives if they have any information to contribute. If you cannot find any information, or if you want to supplement what you have found, phone the company and ask. Just say straightforwardly that you are preparing for an interview and would like answers to a few questions. "Where can I find out more?" is always a good question. Remember that you will be expected to have followed up on any suggestions. Write down some additional questions to have on hand for the interview.

5. **Practise answers to standard interview questions.** Most of the questions you will be asked by interviewers will deal specifically with the requirements of the job and the information you have provided on your résumé. But many interviewers also have some standard questions they ask every candidate, often questions that do not appear very relevant to the job. Some examples of these questions are listed in the section on the employer's perspective on page 253. Or you may be told to "Tell me about yourself," "Tell me a story," or "Describe your ideal job." Of course, you do not want to sound as if you have memorized a script, but it is useful to practise answering these questions so you will not give an unfocused or damaging answer. Everything you say in an interview should support the idea that you can meet the employer's needs in this job. You can find more examples of these questions on Internet job search sites such as monster.ca., along with ideas for answering them.

2 AT THE INTERVIEW

Try to arrive ten minutes early so that you will have enough time to relax but not enough time to get nervous again. Stand up when the interviewer comes into the waiting area to get you or when the receptionist indicates that it is your turn to be interviewed. The interviewer may take the initiative in greeting you; otherwise, put out your hand for a handshake and say, "Hello, I'm...." Wait to be invited to sit down. An interviewer makes an initial decision about a candidate in the first 60 seconds, so a confident manner and eye contact are essential.

There will be a few general remarks, perhaps about traffic or the weather. The interviewer may then ask the "ice-breaker" questions about hobbies and interests mentioned above, or ask you about sports or movies or a current news story. The interviewer's choice of topic probably reveals his or her own interests, so try to be positive even if you are not informed. For example, if the interviewer says, "Who do you think is going to win the Stanley Cup?" avoid making a tactless reply such as "I don't have the slightest idea; I hate hockey." A better choice would be "I don't really follow hockey very closely; I'm a football fan," or "I don't really follow hockey; I'm very involved with music." This will get the conversation going in a direction where you are interested and knowledgeable, without putting down the interviewer. Likewise, if you are offered coffee and you don't want any, try "No thanks, but I'd love a glass of water," rather than "No thanks. Coffee's bad for you."

Then the interviewer will probably spend ten minutes or so talking about you, asking questions about your own perceptions of your strengths and weaknesses, your long-term goals, or what you learned at school and on the job. If you have prepared for this, you should be able to speak fairly fluently and confidently about yourself. Never put yourself down, even as a joke. Don't say, "Well, I haven't done much supervision, except for supervising kids at summer camp." Say, "I've supervised kids at a summer camp." Don't put down teachers, school, jobs, or supervisors, either.

Reinforce your self-confident image by suppressing nervous habits. Don't touch anything on the interviewer's desk. Fold your hands if you find yourself fiddling with your hair or jewellery. Nervous tics are very distracting to the interviewer.

Brief, one-word answers—"Yes," "No," "Five years"—make the interviewer feel like a game-show host. Ideally, you want to develop your answers in a way that shows thought and interest without going on and on. The questions and answers should get into the rhythm of a conversation. After about ten minutes, if things are going well, the interviewer should change the emphasis from you to the job. If this doesn't happen, you had better try harder to make a good impression.

Assuming the interviewer is now describing the job or the organization, your role is to look interested and be prepared to ask questions about anything you don't understand. If something is mentioned that particularly interests you or that you have experience or expertise in, look for an opportunity to point this out. After some discussion about the job, the interviewer will probably ask if you have any questions. He or she will expect you to have done some basic research on the organization. For example, "What is it you manufacture?", "Do you have more than one location?", or "What's the population of the city you're located in?" are not impressive questions. Do not suggest self-serving motives with questions like "Are you planning a move, because I live right around the corner from here?", "Will you be introducing a dental plan?", or "Will I be able to get time off to pursue my training as a biathlete?"

At the end of the interview, the interviewer should tell you when a decision will be made. If he or she doesn't, it is appropriate to ask. Stand up, shake hands, say goodbye, thank the interviewer, and say, "I look forward to hearing from you."

The Employer's Perspective

Here are some typical interview questions:

Why are you applying for this job?
What do you know about this organization?
What have you accomplished recently?
What do you like best about this field of work?
Why did you leave your last job?
What would be your major contribution to this organization?
What was it like working for ... ?
How does your education/experience relate to this job?

What common themes appear in these questions? They are variations on two underlying themes: (1) What kind of person are you? and (2) What can you do for me? Employers are trying to identify competent, motivated, and conscientious people. They are trying to weed out candidates who lack the skills and competence required to do the job. Equally important, they are trying to avoid hiring people who are unmotivated or dishonest, or who are complainers or troublemakers. Those who interview you will be looking at everything you say and do, as soon as you arrive, from the point of view of these important considerations.

3 FOLLOW-UP TO INTERVIEWS

At the end of an interview, it is appropriate for you to ask when a hiring decision will be made. A typical answer would be, "We have interviews scheduled for the rest of this week, and we will be meeting on Monday to make a decision." Sometimes the interviewer will tell you that selected candidates will be asked by a certain date to a second interview, perhaps with someone more senior in the organization.

Write a message to the interviewer or the interview committee, in care of the chairperson, thanking him, her, or them for the opportunity of discussing this job and learning more about the organization. You will find an example in Figure 11.1. Normally this message should be brief and simple. Its purposes are to:

- reinforce the idea that you are polite and respectful of the time and efforts of others;
- underline your interest in this job and organization;
- demonstrate your good communication skills.

A message of this kind always creates a favourable impression. Even if you are not successful in getting this particular job because another candidate was more qualified, there may be other jobs available in this organization, now or in the future. Sometimes first-choice candidates turn down job offers and the other candidates are re-evaluated. An effective follow-up letter then becomes an important factor.

Send the follow-up message through the mail, or e-mail it if you are not certain that it will be delivered in time.

FIGURE 11.1   Sample Follow-up Message

Subject: Today's interview
Date: June/13/2009 1:04:24 PM Pacific Daylight Time

From: Ramon Navaratnam
To: Ms Elinor King eking@goldcoastimp.ca

Dear Ms King

Thank you for the opportunity to discuss the position of management trainee with your company. Gold Coast's plans for expansion in Southeast Asia are very exciting and I believe I could contribute to your success in this area.

I look forward to hearing from you soon.

Ramon Navaratnam

Occasionally, after an interview, you may realize that you did not include some information in your résumé that is very relevant to the job you are applying for. If so, you may wish to fax or e-mail a new résumé or some other documentation to the interviewer(s). Attach a short note expressing your appreciation for the opportunity to discuss this position and explaining that as a result of learning more about the responsibilities, you realize that you have experience or qualifications that were not clearly explained in your résumé. Faxing this material will ensure that a hard copy of your updated résumé is available when your application is being reviewed.

What happens if the "decision date" has passed and you haven't heard from the employer? Many organizations do not inform unsuccessful applicants until the first-choice candidate has accepted the job offer, and this may cause a delay. Sometimes committee members or supervisors have difficulty agreeing on a candidate. Do not be tempted to phone unless the decision is long overdue—at least two weeks after the date you were given. Do not ask why you haven't heard or complain that you need to know because you have other job prospects. Simply ask if a decision has been made regarding the position you applied for. Occasionally you may be told that the job was given to someone else. Of course, it was rude and unprofessional not to write or phone you and let you know if you didn't get the job, but do not draw attention to this. Show that **you** are polite and professional by saying, "Thank you for letting me know."

Follow-up to an Unsuccessful Job Interview

Nobody likes rejection, and if you are interviewed for a job that is later offered to someone else, it is tempting to try to put the experience far behind you. Interviewing is by no means objective or scientific, and sometimes interviewers do make bad decisions. But if you are prepared to be open-minded, you can learn a lot by following up on an unsuccessful interview.

Because some negative feedback from the interviewer is inevitable, most people prefer to discuss an unsuccessful interview in person or over the phone. If the interviewer calls you directly to tell you that you didn't get the job, this is a good opportunity to have the discussion. If you are informed by letter, voice mail, or a call from an assistant, you will have to take the initiative and call the interviewer, saying something like, "I just received your letter (message) letting me know that I didn't get the job as (whatever). I was very interested in this position (or "your organization"), and I'd like an opportunity to discuss what I can do in future to improve my chances." It must be very clear that you are not challenging the decision, but only trying to learn from someone with knowledge and experience.

If you are lucky, the interviewer will simply explain that your qualifications were not as good as those of the successful candidate. If you are particularly interested in this field of work, this will give you insight into what educational and work experience you will need to get the job you want. Or the interviewer may offer some career counselling, pointing out that your skills and interests are better suited to a different kind of job or a different work environment. This may be accurate or way off-base, but at least think it over. The most difficult feedback to accept is criticism of your interview performance, so try to listen without being defensive. If you feel you were misinterpreted, think about why the "real" you did not come across. It is not much use being hard-working, motivated, and knowledgeable if potential employers perceive you as lazy, unenthusiastic, and dumb. What can you do to present yourself more effectively? If more than one interviewer comments on an aspect of your interview manner that puts people off, it is time to work with a teacher or counsellor to change body language or speech habits that are not communicating well. Following are further discussions on how you can learn from an interview.

4 LEARNING FROM THE INTERVIEW

Your first experiences may be somewhat shaky. You may go for interviews and decide part way through that you wouldn't take the job on a bet. But an interview, good or bad, is a learning experience, and if you need practice you should take every opportunity to be interviewed, even if you know you wouldn't accept the job. Here are some valuable things you will get from any interview:

1. **Experience.** The more interviews you have, the less intimidated you will be. You will have a chance to try out different answers to the same question and think of new angles on your interests and experiences.
2. **A chance to observe.** If your career experience is limited, an interview is a good chance to see the inside of a real organization. Look at what people are wearing, especially people in jobs you would like to have. Look at what is on the walls: plans, calendars, charts, and even the type of artwork they have chosen. Try to get a feel for the workplace environment in relation to others you have seen.
3. **Feedback.** If you don't get the job, it is quite all right to ask why, as long as you make it clear that you want to learn from the experience, not challenge the decision. If your qualifications were not adequate for a job you were really interested in, this will help you plan how to upgrade your skills so that eventually you will have a chance at this type of job. If you made some errors in

the interview, it is important to find out so that you won't make them again. Accepting criticism is not easy, and not all criticism is valid, but if you are mature enough to listen to a critique of your interview, you will have a chance to see yourself as others see you and make changes where you feel the real you is not being communicated effectively.

If you gave the interview your best shot but didn't get the job, then you and that employer weren't meant for each other, at least for now. The more experience you have, the better the chances that you'll be ready when the right job comes along.

5 STANDARDIZED TESTS

It has been known for a long time that an interview is not a very reliable way of finding out whether an applicant is the right person for the job. In addition, organizations today are anxious to be able to show that their hiring practices are fair and non-discriminatory. Standardized tests can help solve both these problems. The word "test" may be scary, but, in fact, the only thing you have to fear from these tests is your own nervousness. Unlike the questions in an interview, the questions in employee selection tests have been developed by experts and tested to make sure that they accurately measure the skills or aptitudes the employer is looking for. Scoring is objective and does not discriminate against any group. To prepare to take a job-related test, just relax. Look over the whole test first to get an idea of its format and the specific abilities it is designed to test. Figure 11.2 gives some sample items from typical tests. Read the instructions carefully and then proceed at a steady pace, remembering that no one is expected to get a perfect score.

FIGURE 11.2 Sample Items from a Standardized Test

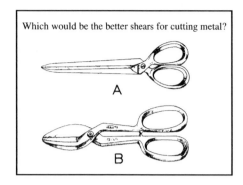
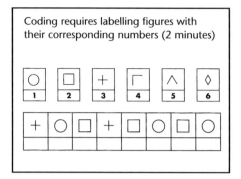

1. Assume the first 2 statements are true. Is the final one:
   (1) true, (2) false, (3) not certain?

   The boy plays baseball. All baseball players wear hats. The boy wears a hat.

2. Paper sells for 21 cents per pad. What will 4 pads cost?

3. How many of the five pairs of items listed below are exact duplicates?
   Nieman, K.M.   Neiman, K.M.
   Thomas, G.K.   Thomas, C.K.
   Hoff, J.P.   Hoff, J.P.
   Pino, L.R.   Pina, L.R.
   Warner, T.S.   Wanner, T.S.

4. PRESENT RESENT—Do these words
   (1) have similar meanings, (2) have contradictory meanings, (3) mean neither the same nor opposite?

5. A train travels 20 feet in 1/5 second. At this same speed, how many feet will it travel in three seconds?

6. When rope is selling at $.10 a foot, how many feet can you buy for sixty cents?

Sources: Top left: *Bennett Mechanical Comprehension Test.*® Copyright © 1942, 1967–70, 1980 by Harcourt Assessment, Inc. Reproduced with permission. All rights reserved. Top right: *Revised Beta Examination: Second Edition.* Copyright © 1946 by Harcourt Assessment, Inc. Reproduced with permission. All rights reserved. Bottom: *Wonderlic Personnel Test.* Copyright © 1992 Wonderlic, Inc. Reprinted by permission.

CHAPTER REVIEW/SELF-TEST

1. The chief purpose of an interview is to
   (a) tell you about the position the employer is trying to fill.
   (b) find out your education and work history.
   (c) introduce you to the objectives of the company.
   (d) determine if your skills and personal qualities meet the employer's needs.

2. Your clothing and personal grooming at the interview should reflect
   (a) your status as a fashion innovator.
   (b) your respect for the job and the interviewer.
   (c) what you would expect to wear to work.
   (d) wealth and a taste for luxury.

3. Which of the following would be the best answer to the question "How long were you employed at Tim Hortons?"
   (a) "Two years; first as part-time counter help, then as a supervisor, and for the last six months I was an assistant manager."
   (b) "It says on my résumé: two years."
   (c) "Two years, but it seemed like two centuries."
   (d) "It says on my résumé two years, but actually it was more like sixteen months."

4. It is important to find out as much as you can about the company or organization before a job interview so that
   (a) you can ask intelligent questions at the end of the interview.
   (b) you can be prepared to show how your qualities and skills would meet the employer's needs.
   (c) you can show your initiative and interest in the job.
   (d) you can do all of the above.

5. The opening questions at an interview are meant to
   (a) fill in time until everyone is settled.
   (b) find out if you meet the objective qualifications for the job.
   (c) evaluate your enthusiasm and people skills.
   (d) satisfy the employer's curiosity about your personal life.

6. Which of the following would be the best answer to the request "Tell me a bit about yourself"?
   (a) "I'm a new graduate from the IT program at Maple Leaf College and I'm really looking forward to the opportunity to do a job where I can use all the things I've been learning about."
   (b) "I grew up in a small town but now I live in Hamilton and I'm hoping to get my own place here."
   (c) "I'm 26 years old and I'm a parent with two boys."
   (d) "Well, I'm just your average kind of guy."

7. At the end of the interview it is appropriate to
   (a) ask if you got the job.
   (b) point out that you are desperate because you have obligations and debts.
   (c) announce that you have been considering a lot of offers because you are in so much demand.
   (d) ask when a decision will be made, if this has not been mentioned before.

8. After the interview you should
   (a) phone the receptionist to see if you can find out how your interview went.
   (b) send a thank you message to the interviewer.
   (c) phone twice a day to see if you got the job.
   (d) do all of the above.

9. If you are turned down for a job after an interview you should
   (a) write or leave a message expressing how unfair this decision was. Don't keep harmful stress bottled up.
   (b) phone and plead with the interviewer to reconsider.
   (c) phone and ask why, so that you can be better prepared for future interviews.
   (d) assume the interviewer was biased or incompetent.

10. If you are asked to take an aptitude test or other standardized test you should
    (a) read the instructions carefully and look over the whole test before you start answering any questions.
    (b) point out that you were not given enough notice to prepare for a test.
    (c) give up if there are any questions you can't answer.
    (d) refuse on principle to waste your time on a useless exercise.

Answers

1.D 2.B 3.A 4.D 5.C 6.A 7.D 8.B 9.C 10.A

EXERCISES

1. Prepare point-form answers to any five of the following questions that are relevant to you:
    (a) Why did you choose your current field of study? How did you go about making your decision?
    (b) Why did you choose your college? Would you choose it again? Why or why not?
    (c) Why did you return to school?
    (d) Why did you change your field of study?
    (e) What do you like best about school? Least?
    (f) What qualities and skills are most important to success in your chosen field of work?
    (g) What contribution have your interests and leisure activities made to your employment skills?
    (h) What have you learned from part-time or summer jobs that will be useful in your future work?

2. Find a partner in the class. Ask your partner five questions that he or she has chosen. Give a constructive critique of the answers from the point of view of a prospective employer. Change roles.

3. Select five job advertisements or job postings for positions you might apply for after graduation. Look for advertisements that give you some information about the company and the job responsibilities. For each position, answer the following questions, in writing, or verbally to a partner:
Why do you want this job?
What do you have to offer this employer?

4. For each of the following questions or requests, list two or three job-related skills or qualities that an interviewer might be trying to identify:
Tell me about yourself.
If you weren't attending this interview right now, where would you most like to be?
Do you think I am a good interviewer?
Describe your ideal job.
Tell me about a book you've read recently.
What is your biggest challenge at work?
Tell me about your biggest achievement.
What would be your idea of a dream vacation?

5. Write a point-form outline of your answer to three of the questions in exercise 4. If you have voice mail, phone yourself and leave your full answer to one of the questions after the tone. Then pick up your message. How did you do?

6. Read the following responses to common interview questions. Place check marks beside those that are satisfactory. Rewrite any that need improvement.

   (a) Interviewer: Where do you see yourself in five years?
   Candidate: Well, I'm really hoping to learn all I can about the travel business so that I can open my own travel agency.

   _____

   _____

   (b) Interviewer: What strengths do you think you would bring to a career as a beautician?
   Candidate: Well, I'm very hard-working and efficient, and I like to focus on a task until it's done.

   _____

   _____

   (c) Interviewer: What was your least favourite subject in school?
   Candidate: I really had a lot of trouble with French. It was a required credit, so I just had to pass it by sheer determination.

   _____

   _____

   (d) Interviewer: Why do you want a career in accounting?
   Candidate: My cousin went to work for an accounting firm, and in a couple of years she had a really great job, with good money and a company car and a cell phone and benefits like that.

   _____

   _____

(e) Interviewer: Tell me about working as a bike courier.
    Candidate: Well, you certainly get to know the city. It gets to be a personal challenge, seeing if you can find the best route, meeting some impossible deadlines.

(f) Interviewer: I see you participated in a volunteer program with a community centre. Why did you get involved in this?
    Candidate: Well, at my school you could get an academic credit for doing so many hours of community service, so I picked this place because it was just up the street from my house.

(g) Interviewer: Have you used this particular computer program before?
    Candidate: Well, only in one of my courses last year.

(h) Interviewer: What is your greatest weakness?
    Candidate: I hate to give up.

PART FOUR

The Report

CHAPTER TWELVE

RESEARCHING THE REPORT

Chapter Objectives

Successful completion of this chapter will enable you to:

1. Define "research."

2. Form main questions and sub-questions to guide your research.

3. Distinguish the types of "information" you require.

4. Identify a range of primary and secondary sources.

5. Design and use a questionnaire.

6. Plan and conduct research interviews.

7. Use secondary research sources.

8. Record information in an efficient, organized manner.

9. Document source material appropriately.

1 INTRODUCTION: WHAT IS RESEARCH?

Why learn about research in a business communication course? The fact is, research means gathering information. Without accurate, complete information, it is impossible to make appropriate decisions in any job. This information may be in people's minds, in a computer, or in books, periodicals, or other documents.

Some jobs require workers to gather information and report it on a regular basis. You may be asked to write progress reports on work being done, incident reports about accidents or unusual occurrences, trip reports about places you have been, or periodic reports that keep management informed about regular operations. The research needed for such reports is usually not extensive. However, sometimes you may find that you are required to conduct research in order to **solve a problem** or **support a new idea.** To solve a problem, you may need to gather a considerable amount of information on which to base your recommendations. To support a new idea, you will need to gather the information that will convince others of its value.

2 BEING SURE OF YOUR PURPOSE

The most important step in research is defining your purpose. If you are pulling together information for a simple informational report—for example, an account of sales figures for the first quarter—your purpose is clear. However, it is not always so easy to set a clear direction in research and stay on course. A very powerful way to focus your search is to form a **main question.** For example, as a municipal employee you might be asked to prepare a report on the city's recycling program. You would need to clarify the main question to be answered. It may be that the city council, under financial pressure, is trying to decide whether the program should be continued. If so, then the following main question would act as a guide in all your research:

Should Windsor's recycling program be continued?

You may then work back from the main question to form a set of **sub-questions** that need to be answered. Here are some examples:

What are the components of the program?
How much does each component cost?
Is this program functioning efficiently?
Are its benefits worth the cost to the taxpayers?

It is important to keep an open mind, changing or adding sub-questions as you learn more.

3 FACTS, OPINIONS, AND LOGICAL CONCLUSIONS

Distinguishing among the three types of "information"—facts, opinions, and logical conclusions—will help you select the information that is relevant to your main question and discard the rest.

Fact

A fact is a statement that can be verified. Verification is a process of objective testing or measurement. Although the word "verify" comes from the Latin word **verus,** meaning true, a fact is not exactly the same as a "true" statement. For example, it may be true that you love your mother, but it is not a fact, according to our definition, because it cannot be measured objectively. When someone says, "Water freezes at 0°C," this statement can be tested, and anyone observing the test will be able to confirm the results. When someone says, "Licorice is delicious," however, no test is available.

Facts are the backbone of any report, so they must be the major object of any research. Facts are necessary in your report because they can be independently verified by your audience, and therefore they make your report credible.

Look at the following statements. Put an X beside the ones that are statements of fact.

1. Toronto is usually hot in the summer.
2. Jennifer Lopez was born in New York in 1970.
3. There are 206 bones in the human skeleton.
4. Johnny Depp is the greatest actor of our time.
5. Ontario, Quebec, New Brunswick, and Nova Scotia joined together in 1867 to form the Dominion of Canada.
6. Canada is a law-abiding, peace-loving country.

If you examine the factual sentences (2, 3, and 5), you will see that one of their common features is the use of words with an **objective** meaning, words that mean the same thing to every reader or listener. Sometimes we use the word "objective" colloquially, to mean "fair" or "balanced."

Non-factual statements use words whose meanings are open to individual interpretation. For example, "Toronto is usually hot in the summer." What does "hot" mean? Some people may be basking contentedly like lizards in the sun when the thermometer hits 32°C. To them, the day is beautifully warm. Others start wilting and refuse to move more than a metre from the air conditioner.

Words like "hot," "great," "delicious," or "law-abiding," while they appear to refer to something "out there" in the real world, really tell us more about the tastes and values of the speaker. This language is **subjective,** because it reflects individual perception rather than something existing in external reality. Objective language is very important to a report because it puts the audience in direct touch with our research, and subjective language makes a report more meaningful and readable.

OBJECTIVE STATEMENT
The main ballroom is 279 m² and has a 9 m ceiling.

SUBJECTIVE STATEMENT
The main ballroom is very open and spacious.

OBJECTIVE STATEMENT
The customer raised his voice and pounded on the counter.

SUBJECTIVE STATEMENT
The customer was angry and upset.

OBJECTIVE STATEMENT
Workers here are asked to handle lawn pesticides without protective clothing.

SUBJECTIVE STATEMENT
Working conditions here are unsafe.

Opinion

An **opinion** is a belief or judgment that may be based on evidence, but is short of proof. Note the opinions below:

She is the best salesperson in the company.
Our meetings are so unproductive!
This company has an incredibly great future.

We make statements like these many times a day. Having a right to one's own opinion is one of the freedoms Canadians cherish. However, opinions are most likely to be seen as valid if they are stated as objectively as possible and supported by facts. Try restating each of the above opinions in more objective language and think of a fact that would give it credibility. Compare your answers with those on page 298.

1. _____

2. _____

3. _____

The opinions you find in your research are respected if they are those of an expert. When you want to get a reliable opinion on a real-estate investment, you ask an experienced broker or bank investment analyst rather than your friends or your hairdresser. Expert opinion is still opinion, but its source gives it credibility because it is assumed that the expert has the necessary information to make an informed judgment. Using expert opinion in reports is discussed in more detail under the heading "Recording and Using Information from Sources" on page 291.

Logical Conclusions

A **logical conclusion** is a judgment based on given data and backed by logic. To reach a logical conclusion, you must proceed from the known to the unknown. Here are some examples of logical conclusions:

1. The man with the Rolex must be rich.
2. Dinosaurs became extinct because of the changes in the earth's climate.
3. The "greenhouse effect" will turn much of the earth into a desert.

Given the price range of Rolex watches, conclusion 1 has probability, but has not been verified. Watches, after all, can be inherited, stolen, or borrowed. An indelicate question directed at the owner can, however, verify his financial status.

Conclusion 2 concerns the past, and although we can make inferences about the past based on given data, we generally cannot arrive at any verifiable conclusions. No eyewitnesses have left us direct evidence about the decline of the dinosaur population, though in this case paleontological discoveries may verify some inferences about the past.

Conclusion 3 predicts the future. We are certainly presented with evidence daily to support this conclusion, but since future events may change the situation, we cannot speak factually about the future.

Every day we go through this process over and over, combining information and experience to come to new conclusions. For example, you might drop in to see your supervisor, Ellen Greer. Ellen's door is open, but she's not at her desk. Her purse is lying on a chair, and there's steam rising from a coffee cup on her desk. Your conclusion? Ellen has just stepped out to some nearby place and intends to return in a minute or two. Of course, the more information you have, the more likely you are to be reasoning correctly. Closer examination of the scene might reveal a note announcing that Ms. Greer has been abducted from her office and is being held hostage by a terrorist organization. Or evidence might suggest that she has been beamed aboard a UFO. A logical conclusion does not have the certainty of fact; it is backed by evidence but has not yet been verified or is not completely verifiable.

Sometimes you will be asked to write reports containing only facts. But at higher levels of management, report writers are often asked to draw conclusions or make recommendations on the basis of the facts that they or others have gathered. There are two ways of doing this: induction and deduction.

Induction is the act of reasoning from particular facts to a general rule or principle. For example, suppose that whenever you visit the home of a friend who owns a cat, you start to sneeze and your eyes begin to water. You suspect the cat may be to blame. You drop in on several other cat owners in the neighbourhood and experience the same symptoms. A visit to a cousin with a new cat also brings on sneezing and tears, although you have been to her house in the past without having these problems. From these events you conclude, "Cats make me sneeze." Because you cannot test this statement with every cat in existence, your conclusion is not verifiable, but it is logical and probable.

When you are doing research for a report, you will usually be using induction. In other words, you will be looking for many pieces of information from which you can draw a logical conclusion (also known as an inference). As in the cat allergy example, your conclusion will be most convincing if you have collected evidence from various sources. In order to arrive at a credible conclusion by induction, you must base your generalization on a sufficiently large sample. Here are two typical **hasty generalizations:**

1. Young people can't handle this kind of work. Our last technician was young, and he was hopeless.

2. The Ford Mustang I've been driving for five years is the best car on the road.

Many more examples would need to be collected before these generalizations would appear logical to your audience.

Deduction is the other way of drawing logical conclusions. A deduction begins with a general statement:

A spider has eight legs.

Then a more specific statement is added that relates to the first one:

A tarantula is a spider.

The logical conclusion is deduced:

A tarantula has eight legs.

Of course, if either statement is not valid, that is, if it is not a fact or a judgment reached inductively, the deduction will not be logically valid. For example:

All birds can fly.
A penguin is a bird.
Therefore, a penguin can fly.

Because the first statement is invalid, the conclusion is also invalid.

The statements must also be sufficiently **relevant** to the conclusion. In the following example, the statements are not sufficiently relevant:

If a car is out of gas, it won't start.
This car won't start.
Therefore, this car is out of gas.

This is called a **non sequitur,** a Latin phrase meaning "It (the conclusion) does not follow (from the preceding statements)."

Look at the following statements. In the space provided, indicate whether each is a fact (F), opinion (O), or logical conclusion (L). The categories may be debatable, so be prepared to defend your choice.

____ 1. Tobacco is physically addictive, making it difficult to quit smoking after only one week.

____ 2. Smoking is a sign of weak character.

____ 3. Approximately half of all regular smokers who begin smoking during adolescence will be killed by tobacco.
____ 4. Smoke breaks down elastin, which is a component of blood vessels.
____ 5. Science has now shown us that smoking leads to many more health problems than just lung cancer.
____ 6. There are at least 43 carcinogens described in cigarette smoke, which include polyaromatic hydrocarbons, heterocyclic hydrocarbons, N-nitrosamines, aromatic amines, aldehydes, volatile carcinogens, inorganic compounds, and radioactive elements.
____ 7. I can quit whenever I want.
____ 8. I am a smoker, and smoking increases the risk of stroke by up to six times, so I am at a much higher risk of stroke than a nonsmoker.
____ 9. Smoking 35 cigarettes or more a day increases the risk of bleeding around the brain, usually due to a ruptured aneurysm, approximately 11 times.
____ 10. Stroke is the leading cause of adult disability.

When you gather information, whether from personal observation, interviews, or printed sources, you will encounter a mix of facts, opinions, and logical conclusions. All can have a place in a report so long as they are used appropriately and not confused with one another. For example, suppose you are investigating the purchase of a new photocopier. As part of your research, you visit an office that has recently installed a particular model you are considering.

You speak to the office manager. "Oh yes," she says. "Great machine. Everyone is very happy with it." Can you write your report confidently asserting the fact that everyone at Company X is very happy with the Smearox LD651? Of course not. The office manager was only stating an **opinion** about the opinions of others. But suppose you questioned her more closely. "Why do you feel everyone is happy with it?" She might present evidence such as absence of complaints or a reduction in spoiled copies entered in the log or compliments the machine has received. This is not enough evidence to verify the statement, but it might support the logical conclusion that the users are happy with the machine. Finally, you might survey the entire staff. If all the questionnaires were returned and all gave the photocopier ten out of ten for satisfactory service, you would be justified in noting this level of satisfaction as a fact. Your reader would appropriately give this fact a lot of weight in a final decision, whereas the unsupported impression of the office manager might not be worth the paper to print it.

Along with the appropriate use of logical conclusions, objective language reduces the appearance of bias in your report and thus makes it more credible. Compare these extracts from two versions of a performance evaluation.

1. X's performance is hindered by his poor work habits and negative attitude. He is not contributing his fair share to the departmental workload, and his output is unsatisfactory.
2. X has been observed arriving late eleven times in the last month. A reminder from his immediate supervisor did not improve his punctuality. On at least three occasions he left for lunch at 12:30 and did not return for the day. On four occasions, claim forms were returned to him as improperly filled out. One of these resulted in financial loss to the company. The average number of claims processed per day has been below the departmental quota and 15–20 percent below the actual average number for the department.

Needless to say, X would not be expected to enjoy reading either assessment of his performance. But the subjective language of the first version would make it easier for him to dismiss the evaluation as merely the opinion of a prejudiced supervisor, the result of a personality clash. The factual content of the second version is not open to this interpretation.

4 PRIMARY AND SECONDARY RESEARCH

Research is usually divided into **primary research** and **secondary research. Primary research** involves gathering information through direct experience, such as experimentation or observation. Questionnaires and interviews are other common forms of primary research. Primary research may take place in a laboratory, factory, office, or any other location.

Secondary research involves locating books, articles, and reports written about a particular topic and selecting the information relevant to your topic. The type of research that will be most important to you will depend on your job, but you should be able to carry out both types of investigation as the need arises. Often you will need to do both in preparing a single report.

Finding the Right Sources

No doubt you have heard that we are living in "The Information Age." Indeed, vast amounts of information are available to us, but we must know how to find what we need. For instance, if you worked for a social agency in a city inhabited by many homeless people, you might be asked to research the situation. Assume your main question is "What should our agency be doing for homeless people?" Think of all the possible ways you might find answers to the following sub-questions:

How many people in our city are homeless?
What social, economic, or health conditions bring about homelessness?
What are their most pressing needs?
What services currently attempt to meet those needs?
How well do these services meet their needs?
What are the costs of these services?
How are other cities dealing with homelessness?

Compare your list with the one on page 298.

A well-developed set of research question helps open your mind to the many potential sources of information, both primary and secondary. A mix of both types of sources may yield the best results, so do not be afraid to mix them. Surveys and interviews with homeless persons and frontline workers would answer many of the important sub-questions listed above, but books and journal articles would be needed to bring a broader perspective to the question, "What social, economic, or health conditions bring about homelessness?" Always think carefully about the blend of sources that would provide the most meaningful and convincing information.

5 PRIMARY RESEARCH: DESIGNING QUESTIONNAIRES

The use of a **questionnaire** requires four steps:
1. Selecting the respondents.
2. Writing the questionnaire.
3. Administering the questionnaire.
4. Tabulating the results.

Selecting the Respondents

Perhaps you want to learn what recreation programs would be attended by people in your community, or how effectively the employees of a company are using their computers, or what the students in your college think their student council should be doing. Ideally you would be able to give a questionnaire to every person in the group you want to learn about. If you have a relatively small group (population), this may be possible. If you want to learn about a large group, you will need to select a sample to represent the whole population. The size of the sample depends on the size of the whole population and the amount of time and resources at your disposal. Don't expect everyone to respond. Rates of response to questionnaires vary with the circumstances, with the nature of the group, and with the way in which the questionnaires are administered, but they tend to be lower than you would expect. Thirty responses from 300 people would be an acceptable sample. Without enough responses, however, you will not have meaningful results.

The least acceptable way to arrive at a sample is by **convenience,** that is, giving the questionnaire to the people who are easiest to contact. A **random sample** will give much more believable results and will also allow you to run statistical tests if you wish. This can easily be accomplished if you know the names of everyone in the population; just draw their names out of a hat. Sampling techniques for more complicated situations can be found in texts on this subject.

Writing the Questionnaire

Three parts are necessary in a questionnaire: introduction, instructions, and questions. The **introduction** must identify the purpose, identify the researcher, discuss privacy, and explain what to do with the questionnaire once it is filled out. It is also wise to explain why a response is important and how much time the questionnaire will take to fill out. Remember that the questionnaire must be short and easy to read.

The **instructions** must be very clear and complete, and at the same time as simple and short as possible. To ensure their effectiveness, try out the questionnaire with a few "guinea pigs" before sending it to your respondents. Of course, the **questions** must be carefully written. There are two main categories of questions: closed and open-ended. Closed questions like the following types are easy to tabulate.

- demographic (age, sex, income, and so on)
- dichotomous (only two possible answers, such as "yes" or "no")
- choice in a list
- rank-order
- scales (for example, ranging from "strongly agree" to "strongly disagree")

Open-ended questions give more freedom to the respondents to say what is meaningful to them, and they add valuable depth to the results, but these questions are more difficult to tabulate.

Administering the Questionnaire

To administer questionnaires, choose a method that would increase the likelihood of getting them back. The response to questionnaires mailed to the general population can be as low as 10 percent. Personal contact with your respondents increases the likelihood of their response. You may be familiar with the methods used in many colleges for student evaluations of teachers and courses. Having clerical staff come to classes with standardized questionnaires ensures a high rate of response and tries to make the process as fair as possible. Another technique is to offer an incentive in the form of payment or a gift, but you must make sure the incentive is not perceived as a reward for answering in a particular way.

Tabulating the Results

Tabulating the answers to closed questions is a straightforward task. The task can be done by hand if your sample is small and the questionnaire short, but using a computer program makes the job much easier and allows you to create tables, graphs, and charts for your report. A statistical program can be used if statistical analysis of the results is desired. Although it takes some effort to learn how to use these programs, the results are worth it.

With open-ended questions, the responses are very individual, and as a result the analysis is more difficult. If you summarize the responses, you must be careful not to introduce your own bias.

6 PRIMARY RESEARCH: PERSONAL AND TELEPHONE INTERVIEWS

Most of the information you require for report writing on the job will not be found in books or periodicals. You will have to get it from other people. Accident reports, sales reports, progress reports, recommendations, minutes—all of these record personal experiences and observations that must be obtained directly from the people involved. As well, it will often be faster and more convenient to get product information and other data by telephone or in person rather than by pursuing written material, provided you are using a reliable source. Good interview techniques will help you obtain useful, relevant information quickly.

Locating Your Sources

First, you must decide whom to interview. Quite often you can find appropriate people by searching the Web. For instance, to learn about addiction counselling available in Toronto, do a search using the keyterms "addiction counselling" and "Toronto." Several counselling services with phone numbers will show up, but remember that not all services have a website, so you will not have a complete list. If you are looking for information about manufactured products, you may want to consult a product and industry directory such as the *Canadian Trade Index*, which is available on the Web. Service industries can be located through your Yellow Pages. Many industries have their own information agencies, such as the Canadian Association of Financial Planners, Canadian Association of Recycling Industries, or the Canadian Institute of Employee Benefit Specialists. Many of these agencies have websites.

Written sources can supply you with a company or association name and telephone number, but a personal inquiry may be necessary to find the specific person who can best supply the information you need. Prepare a well-focused statement of your purpose. The receptionist at Acme Industrial Carpet will not find a statement like "I need some information about industrial carpet" very helpful in deciding who should take your call. Good statements of purpose resemble the following:

"I'm looking for an estimate for framing 35 award certificates."

"I would like some information about establishing the travel incentive program described in the March issue of *Creative Manager*."

"I need some advice on cleaning vertical metal Venetian blinds."

A personal visit to a store or display room may be appropriate, especially if you have little knowledge of the area you are researching and hence only a few vague questions. Choose your informant carefully. Corporate downsizing, especially in the retail area, has reduced the number of well-informed personnel. A tactful opening — "Could you or someone here tell me something about the differences between these two paging systems?" — allows the person you have approached to find a more knowledgeable colleague without losing face. Another choice is, "How can I find out more about the different grades of paper you sell?"

Asking Effective Questions

Once you have located the best source of information, you can begin your research interview. Like all interviews, its success depends on preparation. Know the purpose of your interview, both in general (to select a payroll service for your company; to find out why the cafeteria is losing money), and in particular (to find out what Acu-Cheque Service charges for a biweekly 50-person payroll; to find out how purchasing quantities are established in the cafeteria).

Prepare a list of questions that need to be answered to fulfill this purpose. Then look over your list to ensure that it contains a variety of question types. Some of the most important question types are described below.

Factual

This question expects an objective, verifiable answer.

Examples

Which of your records sold the most copies?
How long has your company been using recycled plastic?

Factual questions provide you with specific data but can be boring for the person being interviewed. Not everyone has a good memory, especially for facts and figures, so information isn't always reliable. Do not use an interview to collect factual information that is already available in print. Ask your source to provide you with or direct you to appropriate written material.

Opinion

This question expects a subjective response or a prediction.

Examples

Why should Canadians be purchasing cruelty-free products?
Does nuclear energy have a future in North America?

People generally enjoy an opportunity to share their opinions, but answers can be time-consuming without adding much useful information. Use these questions for opinions on subjects about which your source has significant experience and expertise. Use follow-up questions (see below) to get supporting data.

Closed

This question restricts the choice of possible answers, often to "yes" or "no" or a brief phrase.

Examples

Which of the two filters performed better?
Are you satisfied with the current production schedule?

Too many closed questions can frustrate your source's wish to have a say in the direction of the interview. You may be allowing your preconceptions to get between you and important information.

Open

This type of question lets your source decide the scope of his or her answer.

Examples

What are some of the things your facility is doing to cut down on waste disposal?
What do you consider Canada's biggest challenge in the 21st century?

Good open questions require careful planning; you want to be thought-provoking without being so open-ended that your source has no idea what you're driving at or where to begin. Use follow-up questions (see below) to move the discussion in a direction that is relevant to your purpose.

Follow-up

A follow-up question can be any of the types described above. Since the purpose of a follow-up question is to clarify, expand, or support a previous answer, the question type is often opposite to the one that went before it.

Examples

Which of your records sold the most copies? (opening question, factual)
Why do you think this record was so popular? (follow-up question, opinion)

Are you satisfied with the current production schedule? (opening question, closed)
Since you're not, how do you plan to speed it up? (follow-up question, open)

Follow-up questions cannot be specifically planned, because the questions you ask will depend on the answers you get to your main questions. Good listening skills and a constant focus on the purpose for which you are gathering information are the keys to asking good follow-up questions.

Make yourself welcome the next time by thanking your source for his or her help. If you eventually make a decision to purchase goods or services from an organization, it is a professional courtesy to mention the person who initially supplied you with information: I was talking to Grace Cooke, your customer service rep, and she convinced me that your lawn maintenance program was really better for our property.

Taking Notes in Person or by Telephone

If you are collecting facts for an informational report—for example, an accident report, sales call report, or minutes—your main task will be to prompt your source when necessary and write quickly enough to record the important information. Write down the time, place, people present, and anything else you can before the meeting or interview begins. Clarify the purpose of your report in your own mind so that you will have some criteria for including or omitting information. Minutes, for example, provide an official record of the decisions of a group; they do not need to reflect the entire process of debate. A sales-call report provides a basis for later follow-up. It should concentrate on recording the needs of the potential customer. Remember to note down names, dates, and similar factual information needed for later reference, since these details will be the first to slip your mind when you write your report.

If you are using a personal source as you would a written one, to contribute to an analytical report, start by preparing a list of headings as you would for taking notes from written sources; they should cover the areas of information you expect to cover. Put these in question form on a sheet or several sheets of paper, leaving space to note the answers. If you don't organize your notes in this way, you may be left with a page of cryptic words and figures that will mean nothing the next day.

Write down the date, time, and name of the person with whom you are speaking before you begin. If you think you detect strong feeling or bias in the speaker, try to determine what objective observations give you this impression: tone of voice, body language, yawns, tears? Put quotation marks in your notes around the exact words of your source. If you are paraphrasing or summarizing, avoid inserting your own opinions or using language that is more loaded or subjective than that of your source.

7 SECONDARY RESEARCH

We often need to find out what has been written by others about our topic. This section discusses the use of libraries, computer databases, and the Internet.

Libraries

Libraries have more information than most people realize. They hold not only books but current and past issues of many newspapers and magazines, government documents, trade journals, academic journals, directories of companies and organizations, videotapes, and more. As the electronic revolution progresses, library research is getting easier and faster. Some libraries maintain Web-based collections that allow you to retrieve information quickly from a variety of sources without moving from your seat at the computer.

Books

The traditional place to start looking for written sources is your library's **catalogue.** Here, all the books in the library are listed by author, title, and subject. Most libraries now have computerized catalogues, allowing you to search very quickly and efficiently. Library staff can teach you in a few minutes how to conduct basic searches.

At most colleges and universities, the computerized catalogues can be accessed by the Internet. If you have Internet access at home, you can conduct searches from there, although you must visit the libraries in person to read most materials. As a student, you may be supplied with a password that allows you online access to full-text articles from publications in various fields.

Whether or not you are using a computer, the **subject catalogue** may be useful if you are starting out with a broad topic. However, be sure you are using the right subject headings. Most academic libraries use the Library of Congress (LC) system to identify and organize their books. The subject catalogue uses the subject headings under which the Library of Congress system organizes information. Unless you are using the same terminology, you will not find the books you are looking for. For instance, you may be accustomed to the term "labour unions" but find that the correct subject heading is "trade unions."

The *Library of Congress Subject Headings*, three large volumes available in the reference section of your library, will help you locate the appropriate subject heading(s) for your topic. Another way to find the appropriate subject headings is to look at the back of the title page in a textbook on your topic; the headings under which the text would be found are listed near the bottom.

With computerized catalogues, **keyword searches** may lead you quickly to appropriate titles if you already have a good idea of what you are looking for. For instance, if you needed to find information on learning disabilities in adults, a keyword search using these terms would be fruitful.

Once you have identified books that interest you in the catalogue, copy down their **call numbers,** for example, D.5.L84. These numbers are your guide to the location of the books in the library.

When you have found the book you want, look at the other titles in the same area, since books on the same subject are shelved together. Remember that your topic may encompass several LC subject areas, so don't assume that every book you need will be on the same shelf. Another very helpful technique is to look in the book for a bibliography or notes referring to the titles and authors of other relevant books. Then consult the **author catalogue** to find these books in your library.

Besides the catalogue, which identifies the holdings of a particular library, there are book indexes that cover all the books published in a given year or on a given topic. Consult your library staff for help with these indexes.

Reference Books and CD-ROMs

Reference works are collections of facts and figures on a range of topics, arranged systematically for easy retrieval. Some common types of reference books are almanacs, handbooks, atlases, dictionaries, and encyclopedias. Within these broad classifications are books covering every aspect of science, technology, history, geography, art, politics, and countless other subjects. Because of their size and scope, reference books have varying levels of accuracy. The best ones cite their sources, so you can check to verify important information for yourself. Your librarian can help you identify the most reliable reference works.

Some reference works are available on CD-ROM. Examples are *The Canadian Encyclopedia*, the *Encyclopedia Britannica*, and *Encarta* (an electronic encyclopedia). You will probably find some of these works on computers in your library, and it is also possible to purchase them for your home computer.

Periodicals

The most updated and specific information on many subjects is published regularly in journals and magazines. Trade journals are published for specific occupations, such as nursing, automotive manufacturing, and hospitality. Scholarly journals cover particular areas of research, for example, the *Canadian Journal of Criminology and Criminal Justice*. Many of these are available via the Internet. Some may be accessed freely just by doing a search on your home computer. For example, a search using the keywords "tool and die trade journal" brings up the *Foundry Trade Journal*.

Colleges and universities provide their students with access to many journal articles via electronic periodical indexes. Educational institutions pay a fee so that their students can search indexes such as *Canadian Business and Current Affairs* or the *Cumulative Index in Nursing and Allied Health*. These indexes are similar to the computerized catalogues and may be available from the same menu. If you work for a large company, you may find that it purchases access to indexes. You will often find the entire article, known as full-text, when you search these indexes; but sometimes you will have to ask the librarian to locate the article in the library or order the article for you.

Your library will have a list of the periodicals it has available in hard copy. A librarian will be able to help you locate periodicals not held by your particular library. More and more periodical articles are appearing in full-text computer databases. In other words, when you locate the name of the article by searching in an index, you may also be able to retrieve the article itself.

Government Documents

A tremendous amount of useful information is available at two important Canadian government websites.

First, **Statistics Canada** (www.statcan.ca) is the place to find up-to-date information about Canadian social, health, manufacturing, and economic trends, as well as many characteristics of our population such as income levels, ethnic origins, or even same-sex unions. For example, in the Community Profiles section, you can look up the number of people in your city aged 20–24 attending school full-time. You will find not only the statistics but also many articles that explain them and their meaning to Canadian society. Whenever you are researching any topic, it pays to do a Statistics Canada site search to see what might be there. If you access the site from home, some material will not be available free in full-text, but college and university libraries purchase access to a special site called E-STAT that provides the information free to students.

The second website is the **Government of Canada Publications** site (http://publications.gc.ca). It is a catalogue of publications by all federal departments as well as parliamentary bills, statutes, and committee reports. Provincial documents are available at the same site. Some documents are available in full-text at the site. Others must be ordered and paid for, but some of these publications may be found in your library.

Computer Databases

A **database** is a central memory in which large amounts of information are stored for the use of anyone who has access to that memory. Many large companies maintain their own databases, which may contain company records, files, customer information, or whatever data the company needs to do business. For reasons of confidentiality, there may be strict limitations on access to databases. In some organizations, certain information may be available to anyone who wants it, while more sensitive information, such as the names and addresses of clients, or employee salaries, may be accessible only to people who have special security clearance.

Another important source of data is the **closed user group.** This is an information-sharing system established by the members of a fairly small, clearly defined group of people who have a special need to share data among themselves. Such a group would be a professional, business, or industrial organization that needs to send and receive specialized material within its own, perhaps widely dispersed, membership.

For instance, a closed user group might be formed by a chain of franchised stores to keep all its franchisees aware of the latest price changes and special promotions. In turn, individual franchisees might use this service to check on delivery dates, to query their accounts payable, or to send messages to other franchisees. Here, the closed user group works both as a form of e-mail and as a highly specialized type of database.

A growing number of commercial databases are operating on a subscription basis. A subscriber can contact the database through a computer at any time to seek whatever information is required. This information may be something as simple as a list of the day's market quotations for a given stock or as complex as a history of that stock's ups and downs over time. There are publicly accessible databases that offer fairly broad ranges of information, and others that specialize in highly technical data on a specialized subject or field.

The Internet

To make use of the World Wide Web for serious research, you need to find the most reliable sources. For instance, current information on the West Nile virus is available at Health Canada's website. A very popular and useful site for those interested in Mars exploration is maintained by NASA. Of course, departments and individuals at universities maintain sites that present and discuss research in many fields, including science, business, and humanities. Many journals are also available online.

Avoid taking information from sites that are not clearly identified or that reflect opinions not based on research. A great benefit of the Internet is that anyone can publish material, but as researchers, we must be selective about what we choose to use.

Another important use of the Internet is to communicate with people. Many sites provide easy e-mail access to the author. News groups and listservs permit discussion of topics by individuals all over the world. Some of the participants are experts; if you have the time, a powerful way to gain insight into a topic is to join a group, follow the discussion, and ask questions.

Searching the Web*

The best websites for reliable information are connected to reputable academic, not-for-profit, and government institutions. The Web address is also called the URL (Uniform Resource Locator), and it contains clues as to the type of site you are visiting. Note the following typical home page address endings:

.com	commercial organization, any country
.edu	educational institution (not always in the ending)
.org	organization (usually not-for-profit group)
.net	varied organizations, usually providing Internet services
.gov	government U.S.
.mil	military U.S.
.ca	Canadian sites; may be any kind of organization but not generally used for commercial sites (there are similar short forms for other countries such as .us for the United States and .au for Australia)

New endings such as ".info" and ".biz" are now being used as well and give an indication of their purpose.

Search engines and directories are the two most common means of finding information on the Internet. For instance, Google is a very popular search engine and Yahoo! is a popular directory. The distinction between these two tools is now somewhat blurred: most search engines have some sort of directory on their home page, and most major directories now offer the option of keyword searches of the Internet.

---

*This section has been adapted from *Nelson Guide to Web Research*, 3/E by Grant Heekman, © 2004. Reprinted with permission of Nelson, a division of Thomson Learning: www.thomsonrights.com. Fax 800-730-2115.

Examples of Useful Search Engines

The following search engines may be successfully used for school and work purposes:

Google	www.google.com
All-the-Web	www.alltheweb.com
AltaVista	www.altavista.com
Lycos	www.lycos.com
Excite	www.excite.com

Directories

Directories allow you to search by categories, although some, such as Yahoo!, can also be used to search by site (like a search engine). Yahoo!, Open Directory, and Looksmart are general-interest directories that provide links, within category menus, to other more specialized directories.

As the Web matures, the major search engines are becoming increasingly similar and interconnected. A single search on a single search engine may give you results from a number of different search engines and directories. For instance, searches on Yahoo! refer to the database of Google.

When you are researching for school or work purposes, **academic directories** may be quite helpful. These directories have databases of scholarly material chosen and organized by librarians or other experts. Here are some examples:

Librarian's Index to the Internet	www.lii.org
Internet Public Library	www.ipl.org
Virtual Library	www.vlib.org
Infomine: Scholarly Internet Resource Collection	http://infomine.ucr.edu

Metasearch Tools

Metasearch tools enable you to conduct research on a number of search engines at the same time. This has the obvious advantage of increasing the size of the database you are consulting. However, remember that the metasearch engines return only a fraction of the results from each search engine they use. A good use of a metasearch is to start a project by comparing kinds of results returned by various search engines. In this way you can get a sense of a particular search engine's suitability for your purposes. Here are three popular metasearch tools:

Metacrawler
www.metacrawler.com
- Uses ten major search engines or any chosen combination of them

DogPile
www.dogpile.com
- Uses 15 general search engines and a number of specialty engines for images, news, audio, and other particular sorts of information

Ask Jeeves
www.askjeeves.com
- Enables you to search by asking questions in natural language
- Responds to the question you posed with a list of similar questions from which you choose one to be answered

Finding Canadian Sites on the Internet

Most of the major search engines and directories allow you to search only for Canadian sites. However, be aware that "Canada only" searches usually find only websites that end in .ca., ignoring Canadian sites that end in .com or other endings. There are some specifically Canadian search tools, including the following:

Canada.com
www.canada.com
- Is a branch of the Southam newspaper chain
- Searches Canadian sites or the entire Web, identifying Canadian sites with the Maple Leaf insignia
- Gathers information from 30 Southam publications

Maple Square
www.maplesquare.com
- Is a Canadian directory offering both keyword searches and browsing by subject categories
- Accepts both formal and natural language searches

National Library of Canada: Canadian Information by Subject
www.nlc-bnc.ca/caninfo/ealpha.htm
- Provides an extensive subject tree containing links to international Internet sites that have Canadian information

Government of Canada Internet Addresses
http://canada.gc.ca/directories/internet_e.html
- Contains links to the websites of Government of Canada organizations
- Offers a comprehensive collection such as the Aboriginal Canada portal and Agriculture and Agri-Food Canada

Techniques for Efficient Searching

To use a search engine, you must construct a sequence of keywords that will produce a manageable number of relevant, useful hits. Some of the most useful techniques for refining your searches are described below. Remember that the ranking of findings is based on frequency and location of keywords, but because search engines have varied rules, the same query will return a different set of responses from every search engine.

Keyword searches. In a basic keyword search in most search engines, you can simply type in your keywords. Always think carefully about what they should be. For instance, if you are researching treatment for depression, you can use the two keywords "depression" and "treatment" to retrieve appropriate sites. You might also want to add the word "studies" if you wish to find highly credible information such as reports of medical research. Often you will discover more specific keywords in the material you find and then you can search using them. For instance, once you discover that depression may be treated by psychotherapy, you can add "psychotherapy" to your keywords in order to narrow your search to this topic.

Phrases. Quotation marks to denote a phrase are recognized by most search engines. For example, you could add the phrase "side effects" to your "depression" and "treatment" keywords. Since some search engines search the whole document, you can put in a part of a sentence that you would be likely to find in the types of articles you want.

Advanced searches and Boolean operators. Some search engines have advanced search options that are more efficient if you find your search is not working well. Advanced searches are normally based on classic Boolean searching in a way that is not obvious to the user. However, it is useful to understand the concept, and some search engines allow you to use Boolean operators yourself.

Boolean operators are the words AND or NOT or OR placed between keywords. A Boolean search for "depression AND treatment" will bring you only sites that contain both these terms, and this search would be useful for finding literature on treatment for depression. A search for "depression OR treatment" will bring you many unhelpful sites that contain only one of the two terms. The Boolean operator "OR" comes in handy when you have keywords that mean the same thing, for instance, "manic depression OR bipolar disorder." Use the Boolean operator NOT if you want to exclude sites that contain material that you don't want. For instance, to avoid sites that discuss drug treatment, you might

HOW TO JUDGE AN INTERNET SOURCE

The quality of information at websites varies greatly. It is essential to review the following points before using a site as a source.

Purpose

- What is the purpose of the site, and is it clearly stated? Beware of someone trying to sell a product or a political viewpoint. Is there an attempt to sway your opinion?
- Since search engines often turn up individual pages within larger sites, look for a link to the home page to see what kind of site it is. The home page should give you more perspective on the purpose of the page you found.

Credibility

- Can you tell who the author is? Look for the credentials of the author(s).
- Where is the information coming from? Is it a university, government, or otherwise reputable organization?
- A bibliography or footnotes should increase your confidence.
- Is contact information for the author(s) or organization included?

Accuracy

- Is factual information consistent with facts found in other sources?
- Are opinions backed up with researched information and presented objectively?

Currency

- When was the article written? When was the site last updated? It is important to find the most current coverage of a topic.

---

type "depression NOT medication." Note that you can't use Boolean terms in Google; use the Advanced Search and you will see how Google does disguised Boolean searching for you.

Summary of Important Search Strategies

1. Directories are a good place to begin if you want to browse and get a sense of what sorts of resources are available on a particular subject. Directory browsing can be useful in narrowing a topic that is overly broad. General

directories lead you to specialized directories in your subject area, which will in turn lead to the actual websites you'll find useful.

2. If you know the precise focus of your subject, it is probably more efficient to go to a search engine. You may want to start with a metasearch engine to help you see which individual search engines will be best for your particular purpose.
3. Be as precise and as specific as possible in your choice of keywords and in the relationships you specify between them. At the beginning of a search, add keywords rather than substituting them. Experienced researchers often employ keyword strings of four or five terms, or even more if necessary. Put the most important of your keywords first. Try two or three different search engines and use different combinations of keywords. Take a quick look at the most relevant site in your first results to see if it contains other keywords you might use.

8 RECORDING AND USING INFORMATION FROM SOURCES

Recording the author, date of publication, and other important information that identifies your source is somewhat tedious. However, taking the time to do it correctly is a mark of professionalism. As you conduct your research for a report, you can make life easier for yourself in two simple ways:

- Create a folder on your computer, and also a paper folder, to hold everything relating to your research.
- For each useful source you find, record the full reference in the documentation style you intend to use. Then your bibliography will be very easy to assemble at the end of your report. Rules for documentation styles are on page 293. Note that the following examples are given in APA (American Psychological Association) style.

Research should concentrate mainly on facts and expert analysis. Here is an example of how a fact should appear in your report:

Canadians consume more energy, per capita, than any other nation in the world (Cullen, 2005).

A book or article may also offer an analysis of the facts. Be sure to indicate that it is someone else's analysis. Here is an example:

Cullen (2005) predicts that only unaffordable gas prices will lead Canadians to reduce their gas consumption.

While expert opinions are not conclusive, they add weight to an argument. Your report will be stronger if you can show that two or more experts hold similar opinions.

If you have conducted a personal interview or attended a presentation, you would indicate a person's opinion in this way:

Terry Cullen, Professor of Environmental Studies, predicts that gas prices will continue to rise dramatically (personal communication, January 21, 2006).

Types of Citations

There are three ways to use information from a source: you can paraphrase, summarize, or quote the original text.

Paraphrase

Paraphrasing means taking the facts and ideas from a source and putting them in your own words. A paraphrase is approximately the same length as the original. Paraphrasing lets you fit the passage smoothly into your report's context.

Original Passage

Although researchers need to further isolate different kinds of fats, we know the incidence of breast cancer is greatly reduced when fat forms less than 15% of the diet.

Example of a Paraphrase

Henning (2005) concludes that eating less than 15% fat significantly reduces the risk of breast cancer. However, this author feels that more research on the effects of different kinds of fat is needed.

Summary

Summarizing also means putting information from another source into your own words, but a summary is always much shorter than the original, giving the main ideas in a condensed form. A summary must not change the essential point that the original author was making.

Example of a Summary

A diet with greatly reduced fat lowers breast cancer risk (Henning, 2005).

Quotation

Quoting means copying a passage exactly as it is printed. Use quotations sparingly because they interrupt the flow of your text. A quotation is a good choice when the original words are particularly expressive.

Example of a Quotation

"Women can do much to protect themselves from breast cancer, but only cleaning up the environment will conquer this ghastly killer" (Henning, 2005, p. 89).

9 DOCUMENTATION OF SOURCES

Identifying your sources makes your reports much more impressive and convincing, and it also allows your readers to find your sources if they find them useful for their own purposes. If your report is informal, you might simply identify your sources in the body of the report, but in a formal report, full documentation is essential.

The most commonly used styles for documenting sources are APA (American Psychological Association) and MLA (Modern Language Association). APA is the standard for business and social sciences. Choose a single style and adhere to it throughout your report. The style rules may seem overly strict, but they lead to a consistent format that everyone can follow.

The terminology used in documenting sources can be confusing, so here are explanations:

- A **source** may be a written piece such as an article, book, or website, or a spoken piece such as a presentation, broadcast, or personal interview.
- The term **"documentation"** refers to the whole process of identifying your sources.
- To **"cite a source"** means to place the last name of the author and the date of publication (APA rules) or page (MLA rules) in the body of your report, usually placed in parentheses (round brackets). See examples below.
- A **reference** is the complete identifying information for a source. All references appear at the end of the report in an alphabetical list, traditionally called the **bibliography.** The title of the bibliography page is **"References"** in APA style, and it is **"Works Cited"** in MLA style. For every piece of information that you take from a source, a citation must appear in the body of the report and a matching reference must appear at the end. This rule stands regardless of whether you are paraphrasing, summarizing, or quoting the source.

It is not realistic to memorize all the picky rules for writing references, so you need to have the guidelines and examples in front of you. Several examples are provided below, and you will find an example bibliography called "References" in the formal report on page 348. You will likely need to go to the Web to get more examples because there are so many types of publications. Here are some very useful sites:

APA Style Websites

Electronic Reference Formats
www.apastyle.org/elecref.html

Using APA Format
owl.english.purdue.edu/handouts/research/r_apa.html

MLA Style Websites

MLA Style
www.mla.org

Using MLA Format
owl.english.purdue.edu/handouts/research/r_mla.html

APA Style Guidelines

In APA style, the last name(s) of the author(s) and the year of publication usually appear together in parentheses (round brackets). However, only the date may be in parentheses, depending on how you structure your sentence. Look at the two variations of this passage:

In a report on Canadian health, Macdonald (2005) described the serious health effects of second-hand smoke on children.

The serious effects of second-hand smoke on children were described in a report on Canadian health (Macdonald, 2005).

If a work does not have an author's name, use the editor's name if there is one. Otherwise, use the title of the work, as in the following example:

Breast implants that have been done strictly for beauty run up costs in the Canadian health-care system when problems arise years later (Cosmetic Surgery Costs All Canadians, 2005).

If you are citing a person whom you have interviewed, the citation would look something like this:

According to a local hospital administrator, there has been a tripling of emergency breast implant removals (M. Renaud, personal communication, June 1, 2006).

Reference Page

The reference page lists full identifying information for all the sources that you have cited. The references are listed alphabetically by author's last name, regardless of the order of the in-text citations. For instance, if you placed the citation (Johnston, 2005) early in your report and another citation (Bikli, 2005) later in your report, the Bikli reference would appear before the Johnston reference in your list. If you examine the citations and reference page in the example formal report that begins on page 338, you will see how the system works.

Examples of References

It is necessary to follow the formatting rules with precision. Note carefully the order of the elements, the punctuation, the italics, and the hanging indent.

BOOK
Hammond, E. (2006). *Cease smoking and feel better*. Toronto: University of Toronto Press.

ELECTRONIC ARTICLE WITH AN AUTHOR
Wong, R. (2005). Tobacco now linked to colon cancer. Retrieved April 4, 2006, from http://www.smokewatch.ca/wong/tobacco.html

ELECTRONIC ARTICLE WITH CORPORATE OR GOVERNMENT AUTHOR
Health Canada. (2005). Cancer and tobacco. Retrieved March 18, 2006, from http://www.hc-sc.gc.ca/english/diseases/cancer.html

JOURNAL
Fiorelli, T. (2006). Marihuana smoke and cancer: Another killer. *Journal of Tobacco Research, 8* (2), 45–56.

NEWSPAPER ARTICLE
Kaur, S.D. (2007, March 11). Inner pollution like smog in body. *Toronto Star*, p. L1.

ARTICLE FROM AN ELECTRONIC DATABASE
Melanson, D. (2005). Massage therapy assists smoking cessation. *Journal of Massage Therapy, 5* (4), 67–78. Retrieved March 14, 2006, from EBSCO database.

Note that an interview is not listed on the reference page.

MLA Style

The rules for MLA style citation are similar to APA rules, but there are a few differences. For instance, instead of the date, the page number always appears in the in-text citation:

In a report on Canadian health, Macdonald (26) described the serious health effects of second-hand smoke on children.

The serious effects of second-hand smoke on children were described in a report on Canadian health (Macdonald 26).

The rules for organizing your bibliography alphabetically are the same as APA rules, but the page is entitled "Works Cited" and the format differs.

BOOK
Hammond, Elizabeth. *Cease Smoking and Feel Better*. Toronto: University of Toronto Press, 2006.

ELECTRONIC ARTICLE WITH AN AUTHOR
Wong, Robert. "Tobacco Now Linked to Colon Cancer." 2005. 4 April 2006. <http://www.smokewatch.ca/wong/tobacco.html>.

ELECTRONIC ARTICLE WITH CORPORATE OR GOVERNMENT AUTHOR
Health Canada. "Cancer and Tobacco." 2005. 18 Mar. 2006. <http://www.hc-sc.gc.ca/english/diseases/cancer.html>.

JOURNAL
Fiorelli, Theresa. "Marihuana Smoke and Cancer: Another Killer." 23 April 2006. *Journal of Tobacco Research*, 8, 45–56.

NEWSPAPER ARTICLE
Kaur, Sat D. "Inner Pollution Like Smog in Body." *Toronto Star* 11 Mar. 2007, L1.

ARTICLE FROM AN ELECTRONIC DATABASE
Melanson, David. "Massage Therapy Assists Smoking Cessation." *Journal of Massage Therapy*, 5, para 15. Ebscohost. 14 Mar. 2006. <http://www.epnet.com>.

PERSONAL INTERVIEW
Asselberg, Sebastian. Personal interview. 24 July 2006.

APA Documentation Exercise

Scenario One

You are doing a report on preventing juvenile crime. You have found relevant information in a book called *Sugar and Life*, written in 2005 by Mahara Kushi. The book was published by St. Martin's Press in Vancouver. Kushi wrote: "At the Bluewater Detention Centre in Calgary, a preliminary experiment with juvenile offenders resulted in 45% fewer infractions and instances of aggressive behaviour among inmates when sugar was removed from their diet."

Paraphrase this information and give the citation as you would in a report. Give the full reference as it would appear in your References list at the end of your paper.

Scenario Two

You are doing a research paper on penguins, and you want to show how extreme their living conditions are. You have read the following in the March 2004 *National Geographic:* "Propelled by crampon-like feet, emperors toboggan toward the Weddel Sea to feed on krill, squid and fish. The distance to a patch of open water in the vast expanse of sea ice varies with the season. In midwinter, birds may have to cross 50 miles of ice at an average cruising speed of half a mile an hour." This article was on pages 66–78 in Volume 183, Issue 7, and was entitled "Emperors of the Ice" by Glenn Oeland and Robert Johnston.

Summarize the passage and give the citation. Give the full reference.

Scenario Three

You are doing a historical research report on patterns in car sales in 1995/96 to compare to the current pattern. In the *Toronto Star* archives, you read the following: "Phil Edmonston, veteran auto consumer activist and former MP, senses a revolution in attitudes toward used cars. He said, 'It used to be in Ontario it was five to six years you'd keep your car. Now it's seven or eight.'" The name of the article is "No shame now in buying used car, critics say" in the March 9, 1996 edition of the paper in section L, page 1. The article was written by Denyse O'Leary.

Paraphrase and give the citation. Give the full reference.

Scenario Four

You are attempting to answer the question "should adolescents be prescribed anti-depressive medication?" You found an article entitled "Medication Blues" at the Canadian Mental Health Association website. There is no author and no date.

The article states, "Today, there is no doubt that the risk of suicide in teens who take anti-depressives is increased by 15%." The URL is www.mentalhealth.ca/news/meds.html. You found the article by searching the site on October 23, 2005.

Introduce and use as a quotation. Give the full reference.

Compare your answers for this exercise with those given below.

Answers to Chapter Questions

Page 270:
1. Arlene has the record for highest sales in the company over the past year. Here are the comparative figures: ...
2. Productivity of meetings needs to be improved. Only one out of five urgent matters was resolved in our last two meetings.
3. We project a doubling of profit in two years. Here are the projects planned: ...

Page 275:
Here are some sources that might provide answers to the sub-questions on homeless people:

interviews with homeless persons and service providers
books and journal articles from the fields of sociology and psychology
directories of associations and community directories
questionnaires sent to social agencies
newspapers
magazines
government documents
reports written for other agencies
the Internet
computer databases

Page 297:
Here are answers to the APA documentation exercise. Your wording of the text may differ somewhat, but the references must be exactly as shown.

1. A study of young offenders at a Calgary detention centre showed that 45% fewer infractions and acts of aggression among inmates occurred when sugar was removed from their diet (Kushi, 2005).
   Kushi, M. (2005). *Sugar and life*. Vancouver: St. Martin's Press.

2. In midwinter, emperor penguins may have to travel up to 50 miles to feed in the ocean, at a speed of only half a mile an hour (Oeland & Johnston, 2004).
   Oeland, G., & Johnston, R. (2004, March). Emperors of the ice. *National Geographic, 183* (7), 66–78.

3. The veteran auto consumer activist Phil Edmonston feels attitudes toward used cars are changing. He says people in Ontario are now keeping their cars seven or eight years instead of five or six (O'Leary, 1996).
   O'Leary, D. (1996, March 9). No shame now in buying used car, critics say. *Toronto Star*, p. L1.

4. The Canadian Mental Health Association (n.d.) has made this strong statement: "Today, there is no doubt that the risk of suicide in teens who take anti-depressives is increased by 15%."
   Canadian Mental Health Association. (n.d.). Medication blues. Retrieved October 23, 2005, from http://www.mentalhealth.ca/news/meds.html

CHAPTER REVIEW/SELF-TEST

1. Gathering expert opinions for a report
   (a) is not worth the trouble because only facts are important in a report.
   (b) will add credibility to your report.
   (c) will make your report too subjective.
   (d) a and c.

2. Deduction is
   (a) thinking on your feet.
   (b) reasoning from a general statement to a specific case.
   (c) not part of our everyday thinking process.
   (d) b and c.

3. Induction is
   (a) the process of arriving at a conclusion based on evidence from a number of sources.
   (b) required only in science and not in business.
   (c) not important in everyday life.
   (d) a and b.

4. Secondary research is
   (a) finding a second person to interview.

(b) searching for information in books, articles, and other written sources.
(c) often done along with primary research.
(d) a and c.
(e) b and c.

5. Which of the following actions is **not** a recommended interviewing technique?
   (a) Write down the time, place, and people present.
   (b) Take notes on a sheet of paper that has room for writing under each question.
   (c) Make sure the interviewee knows your opinion on the subject so he or she can react to it.
   (d) Ensure the wording of your notes reflects the interviewee's opinion rather than your own.

6. Which of the following statements about journals is false?
   (a) Journals are only available on the Internet.
   (b) Trade journals are a good way to keep up with developments in your occupational field.
   (c) College and university libraries pay for access to some journals.
   (d) Scholarly journals provide updates on research into specific fields such as nursing or criminal justice.

7. The Statistics Canada website
   (a) contains statistics such as population size, age, and income levels of Canadian communities.
   (b) contains articles about Canadian society.
   (c) is only valuable for academic research and not for business.
   (d) a and b.
   (e) a and c.

8. An Internet article is more credible if
   (a) it is not trying to sell a product or service.
   (b) the credentials of the author(s) are given.
   (c) the article is at least ten years old.
   (d) there is a bibliography.
   (e) a, b, and d.
   (f) all of the above.

9. Which of the following statements is false?
   (a) A written source is a book, website, or article from which you have taken information or ideas.
   (b) The term "documentation" refers to the whole process of identifying your sources.

(c) A reference is the complete identifying information for a source.
   (d) References are listed in the order in which they appear in the report.

10. Which of the following statements is false?
    (a) It is only necessary to cite the source if you are using an exact quotation.
    (b) Summarizing material from a source is often the best way to incorporate it into a report.
    (c) Quotations should be used sparingly because they interrupt the flow of your report.

Answers
1.B 2.B 3.A 4.E 5.C 6.A 7.D 8.E 9.D 10.A

EXERCISES

1. Consider whether the following statements are facts, opinions, or logical conclusions. There is plenty of room for discussion.
   (a) Good organizational skills are important to a manager.
   (b) Most students who fail do so because they lack intellectual ability.
   (c) If interest rates rise, housing sales will slow down.
   (d) You need a new car.
   (e) The temperature is 26°C right now.
   (f) Any person on welfare is lazy.
   (g) Canada's culture is superior to America's.
   (h) All religions share the same fundamental truths.
   (i) People who need organ transplants greatly outnumber donors.
   (j) Movies aren't much fun without popcorn.

2. Read the following examples of inductive reasoning. Are the generalizations adequately supported by the evidence? Are there any fallacies in the reasoning?
   (a) Visible minorities in this city experience discrimination in housing. Recently, two volunteers took part in a study. One volunteer spoke with a Canadian accent; the other had a Jamaican accent. Both were men. Twenty-six landlords with apartments for rent were telephoned, first by the "Canadian" volunteer, and then by the volunteer with the Jamaican accent. In eleven cases, the first volunteer was invited to view the apartment while the second, phoning ten minutes later, was told the apartment was rented. Then twenty more landlords were called, in the reverse order. This time, in seven instances the first caller was told the apartment was rented, while the "Canadian" volunteer, phoning ten minutes later, was invited to view the apartment.

(b) In a recent study of cancer patients, 71 percent reported periods of depression, sleeplessness, and appetite disturbance up to a year before they were diagnosed. A regular, happy life prevents cancer.

(c) My friend Laura makes $500 a week selling her handmade jewellery at weekend flea markets. Recently I surveyed eight street vendors selling handcrafted jewellery and discovered that the average take was $200 per day. I can earn a lot of money making jewellery and selling it.

(d) In a recent survey of college dropouts, it was discovered that 43 percent had placed in the top group in the reading placement test given to all incoming students. If you are a good reader, you probably won't do well at college.

(e) A survey of 4879 NHL players, past and present, showed that 79 percent of them were born in the first half of the year. Your chances of making the NHL are greater if you were born between January and June.

3. (a) Write a proposal (in memo or e-mail format) to your instructor for a report that you plan to write. Note that, by the time you write a proposal, some research should already be done. Your proposal should contain the following:

INTRODUCTION
Compose an opening paragraph introducing the topic and demonstrating its importance.

RESEARCH QUESTION
State your research question.

SUB-QUESTIONS
List a set of questions that will lead to a comprehensive answer to your main research question.

SOURCES OF INFORMATION
Discuss your success in researching the topic so far and explain what further research remains to be done and how you will go about it.
   Mention any problems you anticipate.
   List three sources you have already found. Use the documentation style required by your instructor. For each source, include a one-paragraph annotation: describe the type of source and its content. Explain how it provides facts, expert opinion, or other material useful to your report.

(b) Write a progress report to your instructor (memo or e-mail) for the same project. Your progress report should answer all of the following questions:

INTRODUCTION
Remind your instructor of your topic. What you have learned about your topic?

RESEARCH QUESTION
Is your main question still the same or have you adjusted it? What sub-questions have evolved? Give a complete list of revised sub-questions.

PROBLEMS
What problems have you had? What surprises have you had, if any?

NEXT STEPS
What is your next step? What help do you need?

SOURCES
What types of sources have you found (books, journals, magazines, handbooks, people, etc.)? What Internet directories, search engines, or electronic indexes have you used and how successful were the searches? What keywords have been successful in your searching?

Add at least three additional sources using the documentation style required by your instructor. For each source, include an annotation.

4. Arrange and conduct an interview with someone who has knowledge that will contribute to a report you plan to write. Write up a brief report of the interview (in memo or e-mail format) for your instructor. Give the name and position of the person and the reason for choosing this person. Describe the steps you took to accomplish the interview, including your first approach, your meeting, and your method of recording the interview. List interview questions and summarize the person's answers to each one. Summarize the information and ideas that this person has given and explain how it will be used in your report. Evaluate the interview process by suggesting what went well and what you would do differently.

5. Prepare a bibliography of at least ten items on one of the following topics. Include at least two books and five online articles. Consult your teacher about the style of documentation he or she prefers.
(a) Role of fluoride in dental health
(b) The meaning of hip-hop
(c) Globalization and Canadian business
(d) Illiteracy in Canada
(e) Manufacturing trends in Canada
(f) Leg-hold traps
(g) Effect of homelessness on children in Canada and the United States
(h) Canadian environmental policy
(i) Carcinogenic effect of common hair dyes
(j) Health problems associated with common pesticides

CHAPTER THIRTEEN

PLANNING THE INFORMAL REPORT

Chapter Objectives

Successful completion of this chapter will enable you to:

1. Define "informational report," "analytical report," and "persuasive report."

2. Understand the basic principles of organizing the introduction, body, and closing of an informal report.

3. Plan and write an informal informational report, such as an incident report, a progress report, a trip report, or a periodic report.

4. Understand the purpose of printed report forms.

5. Plan and write the first draft of an analytical report.

6. Plan and write the first draft of a persuasive report.

1 INTRODUCTION: WHAT IS A REPORT?

A report is an account of what has been learned by any combination of study, observation, experimentation, measurement, experience, and analysis. It organizes the information gained from these activities in a way that is relevant to its purpose and clear and meaningful to the intended audience. In other words, there is no single "correct" report form. This chapter will give you some models to use when selecting and organizing information for any kind of purpose or audience.

Reports generally fall into three main categories: informational, analytical, and persuasive.

- **Informational reports** are written to present information clearly and concisely. Four types of informational reports are discussed in this chapter: incident reports, progress reports, trip reports, and periodic reports.
- **Analytical reports** are written to solve problems. The writer not only presents information but analyzes that information, draws conclusions, and recommends a solution or course of action.
- **Persuasive reports** are similar to analytical reports but are written to convince supervisors, colleagues, or other decision-makers to approve a new idea or practice. Proposals are the most common type of persuasive report.

2 REPORT FORMATS

Short reports (fewer than three pages) are usually written as **informal** reports, that is, as letters or memos. You would use a memo when writing to someone inside your organization, and a letter when reporting to someone outside your organization. In this chapter, all of the sample reports are informal.

A longer and more complex report usually requires a **formal** report format. In Chapter 14, you will find an example of a formal report and a discussion of the sections required in formal reports.

Introducing the Informal Report

The opening section of your report should help your audience interpret the information you are presenting. A good introduction contains at least the following four items: authorization, purpose, methodology/sources, and plan of presentation. The **authorization** tells the reader who requested the preparation of the report, and the **purpose** explains why. The **methodology/sources** section explains how you went about gathering the information in the report and where you found it. This section establishes the credibility of the contents of the report. The **plan of presentation** outlines the contents of the report in order.

Two other items may be appropriate in your introduction: background and recommendations. Give **background** information if your audience will need it to interpret the contents of the report. Make **recommendations** in the introduction of the report if you are using **direct order**, that is, if you have been asked to make recommendations and you do not expect any resistance to them on the part of your audience. If you are using **indirect order** because you think that your audience will regard your recommendations as bad news, or if you are adding recommendations on your own initiative, simply state that recommendations will be found at the end of the report. Figure 13.1 illustrates an appropriate introduction to the report discussed in exercise 1 on page 326.

Organizing the Body of the Report

The body of a report contains findings and supporting details. This is where you will present the information you have gathered relevant to the purpose of your report. Some **objective** models for organizing the body of a report include:

- **Geographic or Spatial.** A report on company hiring could be organized by store location.

FIGURE 13.1 Sample Introduction to an Informal Report

To: Maple Leaf College Student Council
From: Julie Yan Ping Chen, Fashion Retailing Student Representative *JC*
Date: 2008 11 26
Subject: Report on Fashion Boutique

At the request of Travis Ali, President of the Student Council, I have prepared a report on the Fashion Boutique at Maple Leaf College, with particular attention to problems raised by the Fashion Retailing students. Information was gathered from interviews with FRM students, the FRM Course Director, and the College Financial Officer. College policy on laboratory requirements was also researched. The following report presents information on the place of the boutique in the FRM program, the effectiveness of the current boutique as a teaching lab, and the feasibility of expanding the boutique. As a result of my investigation I recommend that the Student Council support the Fashion Retailing Students Association's petition to expand the boutique to be presented to the Board of Governors at their next meeting.

- **Functional.** A performance review of the president of your college might include sections on Government Liaison, Financial Management, Community Outreach, and other functions of the position.
- **Chronological.** A security guard would report an incident that required intervention by describing the events in the order they happened.

These models are objective, or natural, because the order is based on observable or measurable criteria such as location, function, or time. An objective model is often the best choice for a report that presents facts with few opinions or inferences and no recommendations.

Some **subjective** models include:

- **Comparison or Contrast.** The similarities and differences of two software programs could be examined to see which one was better suited to your company's needs.
- **Problem-Solution.** A report on campus traffic congestion would first identify the problem and the factors involved in finding and evaluating a solution, and then analyze the advantages and disadvantages of possible solutions.
- **Priority.** A progress report on IT upgrades on campus would begin with the most urgent requirements and end with "nice-to-have" items.

These models are subjective, or logical, because they depend on the writer's judgment. They are generally appropriate to analytical, persuasive, and problem-solving reports.

Closing the Report

A short informal report can end with the last item of the body. Do not waste your time summarizing what your audience took only a few moments to read in the first place. If you decided to put recommendations at the end of your report, use the heading **Recommendations** and then give a numbered list. If you are sending your report as a letter or memo, you may wish to add a brief action close mentioning appropriate follow-up.

3 PLANNING AND WRITING INFORMAL INFORMATIONAL REPORTS

Incident Reports

Many companies provide printed forms for incident reports (see Figure 13.6). However, if a form is not available, the following organization, also illustrated in Figure 13.2, is appropriate for a memo:

1. **Subject Line:** Clearly tell the reader what event the report describes.
2. **Introduction/Summary:** State briefly what happened, who was involved, when and where the incident took place, and what the significant outcomes were.
3. **Description of the Incident:** Detail exactly what happened, in chronological order, in objective language, and possibly under appropriate subtitles.
4. **Possible Causes:** You may offer your own opinion about possible causes and preventive measures.
5. **Action Taken:** State what you've done to follow up or correct the incident.
6. **Action Required:** List any further steps that need to be taken.

Progress Reports

Progress reports are usually written in memo format to inform management of the progress of a project. However, they are also required for other reasons, as is shown in the sample progress report in letter format in Figure 13.3. In this case a physiotherapist is reporting to the lawyer on the progress of a motor vehicle accident victim. When writing a progress report, describe the work completed, the work in progress, and the work still to be done. In some cases recommendations for adjustments to the project plan may be in order. The following sections are appropriate for a progress report, although variations, such as those illustrated in Figure 13.3, may be used, depending on the type of work being done.

1. **Subject Line:** Identify the document as a progress report, and name the project and dates covered by the report.
2. **Opening/Summary:** Fully identify the project and the people involved. Cover very briefly what has been done, what remains to be done, and whether the work is on time and on budget if applicable.
3. **Work Completed:** Describe what you've accomplished, perhaps including an overall schedule. If you're behind schedule, give the reasons.
4. **Work Underway or Problems Encountered:** Include this section only if needed. It is essential to be objective. Never place blame for a problem on an individual.
5. **Work to Be Completed:** Describe what needs to be done and how long it will take. If you are behind schedule, include a revised version. If you are over budget, request added funding, and justify the costs. Suggest how to solve problems if possible.

FIGURE 13.2 Sample Incident Report

To: Garret Johns, Director
From: Marilyn Solicki, Receptionist MS
Date: February 12, 2007
Subject: Customer Injury at the Front Entrance

At 4:15 PM on February 12, a customer fell on a patch of ice that had accumulated at the front entrance. The customer, whose name is Hema Nalla, was able to walk but went directly by taxi to the emergency department at Queen's Hospital.

Description of Incident
When Ms. Nalla slipped on the ice just outside the front door, she fell on her left elbow. From my desk in the reception area, I was able to reach her quickly and help her to her feet. However, she was in pain and expressed concern as to whether she might have fractured a bone. I offered to call a taxi, which arrived ten minutes later. I waited inside the front doors with her until the taxi arrived. During this time, I took her phone number.

Possible Causes of Ice Accumulation
Since I am stationed near the front door, I am able to see a constant drip of water from the roof onto the concrete when it rains. My assessment is that the accumulation of water from this drip creates a dangerous situation when it has rained and the temperature then drops below the freezing point.

Action Taken
To avoid another accident, I retrieved some salt from the storeroom and spread it on the concrete.

Action Required
I will call Ms. Nalla tomorrow morning to inquire about her condition. The best solution to the problem of ice accumulation, although it occurs rarely, would be to repair the eavestrough over the front entrance.

FIGURE 13.3 Sample Progress Report

GRAAF PHYSIOTHERAPY CLINIC
199 Colborne Street Sarnia ON N7T 6B2

February 3, 2009

Mr. William Nelson
Nelson, Johnson, and Associates
34 Dunlop Street
Barrie ON L4M 3X7

Dear Mr. Nelson

Re: Paul Goldstein's Progress, September 22, 2008 to present

Paul Goldstein has been receiving physiotherapy since September 22, 2008, to rehabilitate injuries sustained in a motor vehicle accident on September 18, 2008. A report of the initial assessment was sent to you on September 26, 2008. Paul has been making satisfactory progress, except for his right shoulder, which appears to need orthopedic intervention. He is now ready to attempt work in a volunteer position similar to his former employment.

<u>Treatment and progress to date</u>
Treatment consisted of education regarding posture, self-mobilization exercises, home stretching, strengthening routines, stabilization, and functional retraining of the trunk, upper extremities, and neck. Paul was shown progressively more home exercises with the use of weights, latex tubing, and "physio ball." I occasionally applied manual techniques to the cervical and thoracic areas to promote flexibility, as well as stimulation to the right shoulder to reduce irritability of the glenohumeral joint.

Paul has shown good motivation and compliance throughout and expressed good insight into his situation. The pain is still a daily presence but much less of a limitation on Paul's functioning. The right shoulder seems to be at a static level, despite regular stretching, and seems to need orthopedic intervention. I wrote a report on November 21, 2008, to his physician, Dr. Angela Harris, regarding my concern about the shoulder. In this report I requested the services of Dr. Carolina Avendano, an orthopedic specialist and well known as an expert on shoulder and knee evaluation and arthroscopy.

Treatment dates
September 25, 27, 29
October 2, 4, 6, 10, 13, 20, 23, 25, 27, 31
November 1, 7, 13, 15, 20, 22, 24, 27, 29
December 1, 4, 6, 8, 15, 16, 20, 27, 29
January 5, 9, 12, 16, 19, 23, 26, 30

Plans
Paul is starting a volunteer assistant position at the Salvation Army shop in February, 2009 for a few hours three times weekly. This will allow him to slowly test the functioning of his upper body and extremities on the job. I will continue to adjust his exercise for at home and at the clinic until we have a thorough impression of his functioning in the shop. Then I will start to plan his discharge, unless the orthopedic intervention necessitates further physiotherapy.

Prognosis
The prognosis for Paul's neck is reasonably good, providing the nerve regeneration progresses as expected. The neck function will never be 100 percent but should allow him to function within his chosen trade. The prognosis for his right shoulder depends on the intervention of a specialist. At the moment, the shoulder has limited function and will not allow Paul to reach a normal level of capacity in his trade.

If you require more information, please call me at 728-1992.

Sincerely

Sebastian Graaf

Sebastian Graaf
M.C.P.A., M.A.A.O.M., C.A.F.C.
Registered Physiotherapist

Trip Reports

A trip report is usually required when an employee is sent on a business trip or to a conference. Even if a report is not required, managers appreciate receiving an account of information or accomplishments that could be helpful to the company. The report should focus on the most important gains made on the trip. An itinerary (chronological account of activities) may be appropriate in some situations and can be included as an attachment. A sample trip report is provided in Figure 13.4.

1. **Subject Line:** Give the name of the conference or destination.
2. **Opening/Summary:** Give the destination, purpose, and the dates of the trip, as well as an overview of the contents of the report. Name the colleagues who went with you.
3. **Background:** It may be appropriate to give background information to establish the context within which the trip occurred, or to elaborate on the purpose of the trip.
4. **Discussion:** Describe the main gains you have made and the benefits to the company. Use appropriate headings.
5. **Closing:** Using an appropriate heading, suggest action to be taken or explain the value of the experience.

Periodic Reports

Periodic reports are those that are produced regularly to keep people informed of some aspect of operations. Some reports, such as accounts of weekly sales, can be produced with very little effort by software designed for keeping track of daily activities. If a company uses point-of-sale software to record inventory, sales, customer profiles, appointment schedules, or other valuable information, reports summarizing any of this information can be produced with a few keystrokes. If you are submitting a report of this kind to a manager, place it in a memo with an appropriate explanation. It is helpful to you and your readers to develop a standard format for such reports. Figure 13.5 is an example of a periodic report.

FIGURE 13.4 Sample Trip Report

MEMORANDUM

To: Marsha Lipman, Sales Manager Date: September 18, 2007
From: Peter Wood, Sales Representative PW
Subject: Western Region Sales Conference

On September 14, I attended the Acme Cleaning Products Ltd. regional conference in Hamilton, Ontario. Don Hammond, Ann Cheung, and I arranged a car pool to keep expenses down. The most interesting and practical parts of the conference were the keynote speech and the packaging display.

President's Message on Environmental Issues
The keynote speech, by Alan Drucker, president, outlined the efforts of the research division to develop more environmentally friendly products. Mr. Drucker cited market research showing that future sales will depend more and more on a reputation for environmental consciousness. He stressed the importance of sales representatives' efforts to make sure retailers understand the company's strength in this area. To increase our product knowledge, he is planning a monthly newsletter that will go to all sales personnel, describing the advances made by our research division.

Packaging Issues
Proposed packaging designs were displayed at lunch with the intention of garnering feedback from sales representatives on their effectiveness. I was pleased to see this move toward more communication between the two departments because our contact with retailers gives us insight into the effects of package designs once the products are on the shelves.

Action Planned
To follow up on the president's keynote speech, I think it would be a good idea to devote ten minutes to discussing the contents of the newsletters at our monthly meetings. If you agree, I would be happy to provide a quick summary of the information and lead the discussion. In addition, I have begun to write down my thoughts on packaging improvements and will submit a memo detailing my suggestions to the director of packaging, with a copy to you.

FIGURE 13.5 Sample Periodic Report

MEMORANDUM

To: Marianne Lightfoot Date: November 1, 2007
cc: Jan Welby
From: Joan Desrosiers, Atlantic Foundations Program (ext. 1631) JD
Subject: Mid-term count for Fundamentals of Mathematics

Here is the mid-term count of the number of students in Fundamentals of Math. We have 363 active students (students who have appeared for at least one test) out of the 407 students enrolled in the course. The 44 students who are not active are all shown on the registrar's list. We can expect that, as usual, many of the inactive students have dropped out of college entirely, but these figures will not be known until next semester. As a matter of interest, I have found at least eight students who were exempted by the College Placement Test but are attending classes.

Figures by Program Area

Program Area	Active	Inactive	Total
Business	162	19	181
Hotel Management	23	4	27
Technology	178	21	199
Total	363	44	407

Please call me if you have any questions.

---

4 PRINTED REPORT FORMS

Many organizations provide **printed forms** for frequently written reports. For instance, incident report forms are standard in hospitals. Although filling out forms may seem an unpleasant chore, they have been designed to save work. The writer does not have to plan a report from scratch, and the people who make use of the reports can quickly locate specific information on the form. Figure 13.6 shows a typical report form.

FIGURE 13.6 Sample Report Form

Incident Report Form

Date: _____ Time of incident: _____

Type of incident: _____
_____

Patient/other involved: _____

Staff involved/witness: _____

Other witness(es): _____

Description of incident: _____
_____
_____

Action taken: Ambulance _____ First aid _____ CPR _____ Other _____

Equipment involved (if any): _____

Signed: _____   _____
          (patient/other)   (staff/witness)

---

5 PLANNING AND WRITING AN ANALYTICAL REPORT

Analytical reports are written to solve problems. The writer must define a problem clearly, present researched information about its causes and possible solutions, analyze the information, draw conclusions, and often make recommendations.

The first step in planning your report is to review your research notes and ensure that all your material is relevant to your **main question.** Focus on the **purpose** of your report, not the **subject** of your report, when deciding what information you will present to your audience. Then prepare an outline, beginning with the main headings you will use in your report. A major difference between a report and an essay or a piece of correspondence is that a report has **conspicuous** organization. Your audience will be guided through your report by headings and perhaps subheadings, which provide an outline of the report's contents. Once you have selected appropriate headings you will want to put them in order. An analytical report generally uses a subjective order such as order of importance or problem-solution. It should be clear to your audience that you have a reason for presenting the information in the order you have chosen; no jumping from minor items to major ones and then back to minor items, for example.

If any section is longer than 150–200 words you may wish to make it easier to read by using subheadings to divide the information into smaller units.

The ways in which information can be divided and combined depend on your audience and purpose, so they are infinitely varied. There are, however, rules for testing the logic and consistency of the divisions you have chosen. Here are some important ones:

1. **Keep all divisions equal in their respective orders.** For example, if you are discussing forest conservation, using New Brunswick, Quebec, and British Columbia as major headings, Northern Ontario cannot be added to this list without violating this rule. You would have to use "Ontario" as the major heading, with Northern and Southern Ontario as subheadings, if required.
2. **Apply a consistent principle of division.** The headings New Brunswick, Quebec, British Columbia, and Ontario indicate a geographical or political principle of division. A major heading like "History of Forest Conservation" would be inconsistent with this principle. It could, however, be used as a subheading under each major heading.
3. **Create subheadings only when discussing more than one topic under a heading.**

    POOR
       1.0   Opponents of Free Trade Deal
       1.1   Labour unions

       2.0   Impact on Selected Industries

    IMPROVED
       1.0   Opponents of Free Trade Deal
       1.1   Labour unions
       1.2   Arts community

       2.0   Impact on Selected Industries

or

✓ IMPROVED
   1.0   Opposition by Labour Unions to Free Trade Deal

   2.0   Impact on Selected Industries

4. **Keep all headings and subheadings grammatically parallel.** Do not mix noun phrases with verb phrases, or sentences with sentence fragments.

✗ POOR
   1.0   Beginning the process

   2.0   To monitor the final product

   3.0   The location of markets

✓ IMPROVED
   1.0   Beginning the process

   2.0   Monitoring the final product

   3.0   Locating markets

In the proposal report on page 321, the four rules of division are effectively integrated.

Once you have created your outline you can write your introduction, which will include a brief overview of your plan of presentation. If you are using the **direct order,** your introduction will also include your recommendations for appropriate action. If you are offering recommendations without having been asked to, or if you feel that your recommendations will be more effective after the audience has read the rest of the report, you should use **indirect order,** placing them at the end in a separate section.

Review your completed report for an objective tone. Remember that a primary purpose of every report is to present information. Support your analysis with facts, which your audience can use to evaluate your conclusions and recommendations.

A sample analytical report is shown in Figure 13.7.

FIGURE 13.7 Sample Analytical Report

To: Members of the Management Committee
From: Filipa Chen, Assistant Building Manager *FC*
Date: 2008 06 06
Subject: Report on Reducing Parking Costs

At the last Management Committee meeting, Building Manager John Soriano reported that providing parking for an estimated 38 new employees would cost the company approximately $12 000 annually. As requested by the committee, the building manager's office has explored ways of reducing this potential expense. Information has been obtained from the Victoria Transport Policy Institute (www.vtpi.org) and the Canadian Urban Transportation Association (www.cutaactu.ca) as well as the General Accounting Manager. This report explains how savings can be realized from reduced parking demand and recommends financial incentives to encourage more efficient commute modes.

Parking Costs
Financial
Currently Acme Systems' building lease includes the use of 153 parking spaces on the surrounding land. On a typical day the lot is filled close to capacity. Additional parking spaces currently cost $438 each per year. Local tax increases and zoning restrictions may cause this price to rise faster than inflation.

Indirect
Acme has no free parking available for visitors, which may discourage some potential clients. Land previously leased for equipment storage was converted to parking space in 2001, requiring the leasing of storage space off-site costing $8300 per year.

Strategies for Reducing Employee Parking Demand
Pay Parking
Ninety-five percent of employees in this community park free at their workplace. Charging Acme employees for the use of a parking space could create problems with recruitment and retention.

Transit Benefits
In the United States, where transit benefits are not taxable, offering free or subsidized transit passes as an employee benefit has been shown to reduce car commute trips by an average of 20%. In Canada, transit benefits are

currently taxable, reducing their potential value to our employees. Efforts are underway to change this policy, so far without success.

Parking Cash Out
This plan offers employees the cash equivalent of the value of a parking space if they use an alternative method of getting to work. Employees can agree to use this method all the time, or a certain number of days per week or month for a smaller percentage of the benefit. This plan has been shown to reduce car commute trips by 20% to 39% in a large number of studies available at the websites mentioned above. Administration typically requires two minutes per employee per month.

Recommendations
Acme Financial Systems should implement a Parking Cash Out plan to encourage employees to use transit, carpools, and other alternatives to single occupant vehicles for getting to work.

To realize the greatest financial benefit from this plan, Acme should request that parking space be shown as a separate line item in its building lease. If the plan succeeds, the need for parking space may drop below the current 153 spaces.

6 PERSUASIVE REPORTS

Although problem-solving reports often have a persuasive element, proposals are written expressly to convince someone to approve or take some kind of action. The greatest challenge to the writer is to anticipate all the questions the reader might ask in order to make a "yes" or "no" decision. The typical headings in a proposal are found in Table 13.1. The headings are matched with questions the reader is expected to ask. They may be varied to suit the content of the report. Figure 13.8 is an example of an effective proposal.

TABLE 13.1 Comparison of Proposed Headings and Reader Questions

Typical Heading	Reader Questions
Proposal	What do you want to do?
Problem Statement	What's the problem and its background? Why do we need to act at all?
Benefits	How will your proposal solve the problem? What's in it for me? For the company?
Project Details (Implementation)	What are the details of your proposal? What are the main tasks for implementing your proposal? What specifically has to be done? How will it all work (the nitty-gritty details)?
Schedule/Deadlines	What's your schedule? What are the details of that schedule?
Evaluation	How will you know if your proposal is successful? How do you plan to evaluate your results?
Other Considerations	What are the potential problems with your proposal? But what if …? How do you plan to overcome these difficulties?
Personnel	Who will be involved in this project?
Qualifications	What are their qualifications? Why should I believe you can do the job?
Cost/Budget	How much will all this cost? Where exactly will the money be spent?
Alternatives Considered	What else did you consider to arrive at your solution? Why did you rule out those solutions? Using what criteria?

Source: From *Business Communication Strategies and Skills*, 4/E by HUSEMAN. © 1996. Reprinted with permission of Nelson, a division of Thomson Learning: www.thomsonrights.com. Fax 800-730-2215.

FIGURE 13.8 Sample Proposal

To: Leona Hutchison, Coordinator, Student Summer Project Fund, Seneca College
From: Tony Bastien, Student, Centre for Individualized Learning TB
Date: March 19, 2007
Subject: Request to Student Summer Project Fund for an interest-free loan as starting capital for a catering service.

Proposal
A survey of job sites in York Region has shown the need for a catering service during the busy construction period from May 15 to August 15. This proposal asks for $2478.38 in a 120-day, non-interest loan to enable three students to set up a business to meet this need.

Problem Statement
During the summer, many seasonal projects—short- and long-term, government and private—are planned for areas not served by established catering companies. Since January 2007, I have been in touch with 37 construction and contracting firms intending to do business in York Region this summer. My survey has identified 18 sites starting up from May 15 that are still without catering contracts, and they are interested in having students supply this need. Each site employs an average of 23 workers. The total workforce is 415. More sites may become available.

Area of Operation
The area of York Region surveyed is bounded by Steeles Avenue in the south, Hwy. 404 in the west, Hwy. 48 in the east, and the Stouffville Side Road in the north.

Employees
There will be three employees: one driver/salesman and two cooks. I have delivered food products in the past and have an extensive background in retail sales.

Gerry Big Canoe is a student in George Brown College's chef program. He has worked several summers at the Royal York Hotel.

John McBeak is a business management student at the University of Toronto. Last summer he worked for a catering company.

(continued)

Operating Procedures
Cooks:
Sunday to Thursday, food will be prepared during the day and refrigerated overnight. (We live together, so we will cook on our own stove and use our spare refrigerator.)
Monday to Friday, at 8:00 AM, one cook will pick up fresh produce and supplies at the wholesaler. At 10:00 AM, the other employee will make the bank deposit.

Driver:
 8:00 AM Heat up food in microwave and place everything in coolers in the van.
10:00 AM Leave for first call.
10:45 AM Make first call on route designed to cover 18 sites in 3 1/2 hours (about
    average for a caterer).
 2:15 PM Finish lunch route and begin return calls to larger sites.
 4:00 PM Fuel van and return home.

Supplies and Equipment Needed
—Van (Econo-Van leased for $340.00 per month)
—Coffee urn (100-cup; works on 12 volt auto system $299.39)
—Coolers (two for hot food, one for cold food, one for cold drinks $213.79)
—Stove
—Microwave oven—at home
—Refrigerator
—Food supplies: A friend in the catering business told me that with 415 potential customers, I'd have to prepare 50 hot lunches, 150 hot sandwiches, 100 cold sandwiches, 200 cups of coffee, and 200 cold drinks per day, not including snacks.

Table A
Operating Costs and Income
Breakdown of Food Costs:

	Cost	Retail	Cost	Retail	Gross Profit
Hot lunch (50)	$1.95	$2.75	$ 97.50	$137.50	$ 40.00
Hot sandwich (150)	1.30	1.80	195.00	270.00	75.00
Cold sandwich (100)	1.00	1.40	100.00	140.00	40.00
Coffee (200)	.20	.40	40.00	80.00	40.00
Cold drinks (200)	.45	.55	90.00	110.00	20.00
			$522.50	$737.50	$215.00

Table B
Detailed Budget

Start-up Costs

Fixed Costs:
—Coolers 4 at $49.95 + tax = $199.80 + $13.99	$ 213.79	
—Van @ $340/pm × 3 mo. = $1020.00: Deposit on van	340.00	
—Coffee urn $279.99 + 19.60 tax	299.59	
—Insurance for 3 months	<u>162.50</u>	
	$1015.88	$1015.88

Operating Costs (First Week):
—Gasoline $30.00/day × 5	$150.00	
—Salaries $120.00/day ($5.00/h × 8h/d × 3) × 5	<u>600.00</u>	
	$750.00	750.00

Food Preparation Cost (First Week):
—See Table A $522.50 × 5	$2612.50	<u>2612.50</u>
		4378.38
Less own start-up capital		<u>1900.00</u>
Total amount requested		$2478.38

Table C

Total Projected Costs and Income (12 weeks or 60 days):
—Fixed costs from Table B	$ 1 015.88
—Van for two remaining months	680.00
—Operating costs (Table B) $750.00 × 12	9 000.00
—Food costs (Table B) $2612.50 × 12	<u>31 350.00</u>
	42 045.88
—Retail income (Table A) $737.50/d × 60	<u>44 250.00</u>
—Net profit	$ 2 204.12

<u>Conclusion</u>
Table A shows that the daily gross profit will be about $215.00.

(continued)

Table B outlines the fixed start-up costs plus the first week's costs for operating and food preparation. These figures form the basis for the funds requested under the student assistance program.

Table C indicates that when all is said and done at the end of the summer, there should be a healthy net profit for the three partners to divide.

As you can see, this gives us considerable leeway if the profits are not as good as anticipated. On the other hand, we may do better.

The fact that we are putting up 43.4 percent of the start-up capital indicates that we are serious in our planning. We are considering a partnership in this area after we graduate, with hopes of hiring students to fill future summer needs.

I would be glad to discuss this proposal with you at any time. I would also appreciate any suggestions.

CHAPTER REVIEW/SELF-TEST

1. A report is considered **informal** if
   (a) you don't worry too much about planning or presentation.
   (b) it has no more than three pages and is presented as a letter or memo.
   (c) it provides information but does not draw conclusions or make recommendations.
   (d) its contents have not yet been completely researched or verified.

2. The **introduction** to an informal report should contain at least
   (a) the authorization, purpose, methodology, and plan of presentation of the report.
   (b) a summary of the report and any recommendations you wish to make.
   (c) a brief statement of your personal views on the subject of the report.
   (d) three important things that the reader will learn by reading the report.

3. The most important factor in deciding where to place any recommendations you will be making in your report is
   (a) visual impact on the page.
   (b) the length of your report.
   (c) maintaining the reader's interest in reading the entire report.
   (d) whether you expect your audience to respond positively or negatively to your recommendation(s).

4. The **body** of an informal report
   (a) follows an invariable format.
   (b) contains five sections: who? what? where? when? why?
   (c) presents findings and important details in a meaningful order.
   (d) does all of the above.

5. The **conclusion** of an informal report should not require
   (a) a summary.
   (b) recommendations.
   (c) any suggestions regarding appropriate follow-up.
   (d) any of the above.

6. Incident reports, progress reports, trip reports, and periodic reports
   (a) are examples of informal informational reports.
   (b) are usually organized in an objective or natural order.
   (c) can often be done on a printed form or computer template.
   (d) all of the above.

7. Analytical reports
   (a) present opinions and logical conclusions rather than facts.
   (b) can often be done on a printed form or computer template.
   (c) use facts and logic to define a problem and present possible solutions.
   (d) always use the direct order.

8. Headings for the sections of your report
   (a) are optional.
   (b) should give the reader an outline of the contents of the report.
   (c) can be limited to three: Introduction, Body, Conclusion.
   (d) should be general so as not to give away too much information.

9. Which of the following is **not** a rule for dividing your report under appropriate headings?
   (a) Apply a consistent principle of division.
   (b) Keep all headings and subheadings grammatically parallel.
   (c) Keep all divisions in order.
   (d) Create a subheading only when discussing at least two items under a main heading.

10. Unlike an analytical report, a proposal
    (a) is explicitly committed to persuading the audience to make a specific decision.
    (b) uses facts and logic to support its conclusions.
    (c) makes a recommendation.
    (d) can be presented as a memo or a letter.

Answers

1. B 2. A 3. D 4. C 5. A 6. D 7. C 8. B 9. C 10. A

EXERCISES

1. Your college operates a fashion boutique on the ground floor of the main college building. It is used as a teaching lab by the fashion program. All fashion students are required to work in the boutique as part of their course requirements. Students have complained for a number of years that the boutique is too small. You have been asked to investigate problems with the boutique and to prepare a report for the student council. Here are the notes, in no particular order, that you gathered from students, faculty, and the college's Financial Officer.
   - The boutique stocks women's clothing and accessories.
   - The boutique has an area of 22 m^2.
   - All students in the fashion program must work in the boutique.
   - Boutique receipts are given for all purchases.
   - There is no storage area in the boutique.
   - College policy allows 1 m^2 of lab space for each student.
   - Forty-seven percent of the college's students are women.
   - If more space were available, the boutique manager would like to involve marketing and accounting students in running the boutique.
   - There are 237 fashion students.
   - Students consistently give working at the boutique high marks in their course evaluation.
   - The main customers of the boutique are members of the college support staff.
   - There are two change rooms in the boutique.
   - Sixteen fashion students are men.
   - Taking over the lounge next door would double the area of the boutique.
   - Customer traffic is highest between noon and 2 PM.
   - Expanding the boutique would cost $113 000.
   - The lounge is now a lunch spot and make-out zone.
   - Spending per student in the fashion program is the third-lowest in the college.
   - With increasing enrollment, leisure space for students is declining in absolute terms.
   - Everything in the boutique is sold at reduced prices at the end of the school year.

Create an outline for your report, grouping your information in a logical order. Eliminate any irrelevant material. Write a 150-word introduction containing your principal recommendation(s).

2. You are the Assistant Director of Building Services at your college. Currently the college spends $473 000 per year to have its garbage collected. You have been asked by your supervisor, Fatima Khan, to think about ways this expense could be reduced. After looking at information about composting on Internet sites, checking with local government sources, and talking to a manager at another college, which has a composting program in place, you believe that composting could be part of the solution at your college.
   (a) Create an outline of appropriate headings and subheadings for a two-page report on establishing a composting program at your college.
   (b) Write the introduction to your report. Include at least two recommendations.

3. Write a periodic report covering your school-related activities for the past week.

4. You have received the following e-mail from your supervisor at Helping Hands Placement Service.

> Subject: Search for new marketing ideas
> From: Robert Cirillo, Director
> Date: Wed, 31 May 2006 14:01-0400
> To: emailuser <emailuser@helphands.ca>
>
> According to the front page of today's business section, 47% of employed Canadians work in businesses with fewer than 100 employees. Is there some way Helping Hands could do a better job finding clients in this sector? Currently our agency places almost all its temporary workers in medium and large firms. E-mail me with your thoughts on this before our meeting on Friday.
>
> Robert Cirillo, Director

Which of the following short reports would be a better response to this message? Explain your choice.

(a)

> I was very happy to get your request as I have been thinking about this issue myself. I decided to talk to some of the small businesses which do use our services and ask them to tell me what attracted them to Helping Hands. Typically a business uses a temp agency when they foresee that workers will only be needed for a short-term project or when an absent worker needs to be replaced until he or she returns. Companies do not want to make a long-term commitment in these circumstances. Because large companies experience these situations all the time, it is natural for them to deal regularly with a placement agency. A small business may have this situation arise very rarely. I think that in these situations they are more likely to try to make do by double-tasking workers who are already there.
>
> But hiring a temporary worker is really a more efficient and cost-saving choice. Of course, if we say this it looks as though we are just trying to get more business. I think we should ask some of the small businesses who do use Helping Hands to offer testimonials in an advertising campaign. In my opinion a small business owner would be more likely to take the word of another small business owner.

(b)

> In response to your message I looked in the company database to find the names of two small businesses which have used Helping Hands in the last year, and contacted the managers to ask why they used our service. Both Sherene Hariprasad at Fitness Basics and Tim Chan at Acme Systems explained that they had previously worked at larger companies which had hired workers from Helping Hands. They knew from that experience that hiring a temporary worker is more efficient and cost-saving than trying to get regular workers to take on extra tasks. If we could get this word out to more small businesses we could build our client base. An advertising campaign using the experiences of real small business owners could help us do this. Both Hariprasad and Chan were willing to be involved. Let me know if you would like me to follow up on this for Friday's meeting.

5. You are the Assistant Director of Communications for the Canadian Olympic Hall of Fame. To publicize the Hall of Fame, the director wants to place an advertising supplement celebrating the centennial of Canada's first Olympic gold medal in *Maclean's*. The magazine publisher is interested in the concept, and you have been asked to prepare a proposal for her. Which of the following would be a better outline? Explain your choice.

(a) I. The history of the Canadian Olympic movement
   A. The founding of the modern Olympic Games
   B. Canada participates for the first time
   C. Olympic Games in Canada
   D. Who are Canada's Olympic medalists?
II. The Hall of Fame magazine insert
   A. Historic events to be included
   B. Athletes to be featured
   C. Interviews and pictures
   D. Production schedule
III. Pros and Cons of magazine insert

(b) I. Introduction: Why is this insert a good idea?
   A. Insert would make money for magazine
   B. Insert would raise Olympic profile
   C. Insert would attract visitors to Hall of Fame
II. Insert Description
   A. High points of Canada's Olympic history
   B. Interviews with Olympic medalists
   C. Pictures of Hall of Fame exhibits
   D. Advertisements
III. Production Plan
   A. Organizational overview
   B. Timeline
   C. Magazine's responsibilities
IV. Next Steps

6. Create an outline suitable for a report on the factors that influenced your choice of college program.

7. Indicate which report plan would be most effective for the following situations: indirect order (recommendations at the end) or direct order (recommendations within the introduction).
   (a) The audience reading report has no power to implement or change recommendations—report simply keeps the audience informed. _____
   (b) The report's conclusions are good news. _____
   (c) The report's recommendations are very technical and complex. _____
   (d) The supporting data are very technical and complex. _____
   (e) The audience has a strong bias. _____
   (f) The audience is pressed for time. _____
   (g) The recommendations are significantly different from what was originally suggested. _____
   (h) The conclusions of the report are bad news. _____

CHAPTER FOURTEEN

PRESENTING THE FORMAL REPORT

Chapter Objectives

Successful completion of this chapter will enable you to:

1. Understand the importance of attractive and readable report presentation.

2. Prepare and organize the structure and all the parts of a formal report.

3. Select graphic elements to give your report clarity and impact.

4. Edit the final draft of a report.

1 INTRODUCTION: THE IMPORTANCE OF PRESENTATION

On the job, making a report attractive and readable is as important as researching and writing it. Your professionalism is reflected in the appearance of your report. Its readability has a strong influence on how its content will be received. Fortunately, high-quality presentation is becoming much easier with the word processing programs available today. These programs give you the power to rearrange and edit material to achieve optimal organization and to produce elements such as a table of contents automatically. They also allow you to easily create and incorporate visual elements such as tables, graphs, and charts. Some programs even offer you templates for various types of reports, although pre-prepared formats may not suit your needs. To help you navigate your way through the myriad of options, this chapter provides you with an understanding of the required parts of a formal report, the most common types of illustrations, and the final editing process. Once you are in the work force, you will need to meet the basic requirements for the type of report you are presenting and be alert to the specific conventions of your particular organization and field.

2 STRUCTURE OF A FORMAL REPORT

In Chapter 13, you studied the structure of three types of informal reports: informational, analytical, and persuasive. The formal report format is usually used for longer analytical and persuasive documents. An additional characteristic of a formal report is greater formality of language. Formal and informal reports share the same basic structure, but a formal report has additional parts. Table 14.1 shows the order of a formal report. A report may follow the **indirect order,** with the recommendations at the end of the body. The sample report in Figure 14.3 follows this order.

A report may follow **direct order,** as noted in Table 14.1, if the audience is likely to welcome the recommendations; in this case the recommendations follow the introduction. The overall organization you see here is preferred in many business settings, but the order of the various parts may vary.

Parts of a Formal Report

In organizing your formal report, it is important to understand the purpose of each part. As noted above, the order of these parts varies, but the purpose of each remains the same.

TABLE 14.1 Order of Formal Reports

Transmittal letter or memo may be placed at the front of the report as a separate piece attached with a paper clip or may be placed inside the report after the Table of Contents. If you are sending reports electronically, it makes sense to use an e-mail message as the transmittal document.

Cover may be needed to protect the report.

Title page

Table of Contents

List of Figures (tables, graphs, charts, diagrams, etc.) may be on the same page as the Table of Contents.

Summary, abstract, or executive summary. Note that in some companies, the convention may be to place the summary after the title page.

Introduction: Always begin a new page for the Introduction. If you are using direct order, place the recommendations after the Introduction.

Body of report under appropriate headings. Do not use the word "Body" as a heading.

Conclusions

Recommendations: If you are using indirect order, place the recommendations here.

References

Appendices

Transmittal Memo or Letter

If you are delivering a formal report to someone within the organization, attach a memo of transmittal (see Figure 14.1). This memo identifies the report, gives the reader any information that will help him or her understand and use the report more effectively, and closes with a goodwill message. If a report is prepared for someone outside your organization, use letter format. If the report is sent electronically, the e-mail message becomes the transmittal document. The letter of transmittal in Figure 14.2 illustrates the order in which information is usually presented. Follow these guidelines for letters or memos:

1. Ensure that the opening sentence gets right to the point: "Here is your report."
2. Identify the subject and purpose of the report, and the person or group that requested it.

FIGURE 14.1 Sample Memo of Transmittal

CANADIAN ADVENTURE TOURS INTER-OFFICE MEMO

TO: Elizabeth Giardino, Director DATE: September 24, 2007

TW WB JG

FROM: Tracy Wright, Werner Blotz, Jean Gauthier
SUBJECT: European Marketing Strategy Report

Here is the report that you requested in your memo of September 3, 2007. This report gathers data on travel patterns in five European countries, presents relevant statistics from the 2006 survey carried out by Foreign Affairs, and surveys the competition in key European markets.

The Recommendation section outlines a marketing strategy consistent with this information. We see many exciting possibilities in selling Canadian Adventure Tours in Europe and would be pleased to discuss the report with you in further detail if you decide to pursue this opportunity.

---

3. Summarize the contents or findings of the report. This section is **optional.**
4. Provide suggestions for applying and following up on the information in the report.
5. Express appreciation for the responsibility entrusted to you and a willingness to be involved in any further way that is appropriate.

Although the letter in Figure 14.2 accompanies a formal report, its tone is more personal than the report itself. You-centred language establishes a friendly, helpful tone.

Cover

A cover protects a report. This may be important if the report is intended to be kept for a long time or handled by many readers. A cover also gives a report a more formal, significant appearance. However, the extra bulk of a cover may not always be appropriate. If you decide to add a cover, choose a weight that is consistent with the length of the report. A clear plastic folder with a slip-on plastic spine would be your best choice for a five-to-ten-page report. A longer, thicker report may be best presented with plastic binding or in a binder. The title of the report may appear on

FIGURE 14.2 Sample Letter of Transmittal

Mountain Marketing Services
Vancouver BC   V4N 1R2

March 4, 2007

Mr. Bill Nelson
President, Ambrosia Herbal Products
83 Luna Avenue
Burnaby BC   V2A 4X6

Dear Mr. Nelson

Here is the marketing analysis that you requested in our meeting on February 1, 2007. This report surveys product sales in five health food chains and compares your product line with the leading sellers in ten areas.

The concluding section of the report makes suggestions regarding new product lines that your firm may wish to explore.

Please phone me at 664-9823, extension 321, if you have any questions regarding the analysis or recommendations. It has been a pleasure working with Ambrosia Herbal Products. If you would like to carry out further investigation of the product lines suggested in the report, Mountain Marketing Services would be happy to assist you in your research.

Sincerely

Janice Schulze

Janice Schulze, Research Director

an adhesive label or through a cut-out window in the cover. Remember that adding a cover may subtract one to two centimetres from the left margin of the page. Leave enough space on the left side of the report's pages so that a margin remains when the report is opened normally, without bending.

Title Page

The title page of a report provides information that helps your audience read and refer to your report. Information required on a title page includes:

1. name of the organization, committee, or person submitting the report
2. name of the organization, committee, or person receiving the report
3. title of the report
4. authors' names if different from the person(s) submitting the report
5. date of submission of the report

The appearance of the title page creates an important first impression. It requires attractive spacing and balance. The most important information should stand out from the rest of the page. Occasionally it may be acceptable to use special fonts on a title page, but do not use Christmas trees, skulls and crossbones, or any other fancy graphics provided by your software. See the example in Figure 14.3.

Table of Contents

This feature gives the audience an overview of the report before it is read and helps the reader locate specific parts of the report for later reference. After the report is in final form, with the pages numbered, you can prepare a table of contents. Word processing programs allow you to create a table of contents automatically from the headings and subheadings in the finished report. If you do not use the automatic feature, be sure the table reproduces exactly the headings and subheadings in the text. If you number or letter the headings, show these numbers or letters as they appear in the report. A table of contents is shown in the formal report in Figure 14.3.

Summary

In a formal report, the **summary** is always separate from the **introduction.** Other names for the summary are **abstract, synopsis,** and **executive summary.** A summary contains a brief statement of the purpose of the report, the conclusions, recommendations, and, sometimes, major findings. Normally, the summary is one paragraph long. However, very long reports often have an executive summary, which is divided into sections and may be up to two pages long. Although a summary may seem unnecessary because all the information is in the report, remember that readers will look to your summary for the "whole story in a nutshell" and may read no further. Don't forget that the summary is always written last but is usually read first. Leave plenty of time to write it because quality counts in first impressions.

Introduction

This section briefly sets the report in context for the reader and states the purpose of the report. It may also give a description of your method of research and an overview of your report's scope and plan of development. If the circumstances leading up to the report are complex, you may want to discuss them in a separate background section. Otherwise, put this information in your introduction. A discussion of the secondary sources relating to the topic, often called a literature review, may be included after the introduction as well.

Body

As you do in an informal report, present detailed information under appropriate headings. The language should be logical, factual, and objective. Never use the word "Body" as a heading. All of the information, whether it is summarized, paraphrased, or quoted directly, must be cited according to one of the formal styles described in Chapter 12.

Conclusions

As occurs in an informal report, conclusions must refer directly to the problem and proceed logically from the information presented in the body. No new information may be included at this point. The conclusions should be concise and may be numbered or in paragraphs.

Recommendations

Recommendations are succinct statements of what should be done about a problem. Connected closely to the conclusions, they must proceed logically from the information presented in the body of the report. If you provide two or more recommendations, number them. If you decide to use direct order, place the recommendations immediately after the introduction.

Reference List

This section documents your sources of information. Its exact title will depend on which style of documentation you are following. Review the information in Chapter 12 before preparing your reference list. It should begin on a new page and be included in your table of contents.

Appendix

An **appendix** (plural, **appendices**) is a section at the end of a report containing information that supports the conclusions and/or recommendations in the report. Some writers use appendices only for information of secondary importance that is not directly part of the logical structure of the report. For example, if the report summarized and interpreted the results of a questionnaire, the questionnaire might be reproduced in an appendix. Or appendices might be attached to provide copies of letters, reports, and other documents mentioned in the main report, or to reproduce data from secondary sources used in calculations or other intermediate steps.

Other writers use appendices to hold **all** detailed data, primary as well as secondary. The main report is thus limited to an analysis of the data and any conclusions or recommendations that can be made.

The needs and expectations of your audience should guide you in deciding where to place data such as tables of figures, statistical tables, calculations, charts, and graphs. On the one hand, quantities of detailed information, especially "number-crunching," can overwhelm the reader and break up the logical flow of the written part of the report. On the other hand, if all the documentation is relegated to the back, the remaining text may lack persuasiveness and credibility. An interested reader may resent having to flip to the back every time he or she wishes to see the hard evidence on which your analysis is based. The best solution probably is to experiment with the placement of your supporting data to judge the effect on your audience. A computer makes it easy to move material around until you are satisfied with the balance of text and detailed data.

The Whole Report

Once you have prepared all the parts of the report, it is time to put them together. Figure 14.3 is an example of a formal report written by a secretary who was given the job of research officer in a manufacturing company. Note that this full report matches the presentation shown in Figure 8.1 on page 179. It follows the indirect order and overall organization shown in Table 14.1. The citations and documentation follow APA style, which is described in Chapter 12 on page 294. Since much of the research for this report was done on the Internet, you will find several examples of website references.

FIGURE 14.3 Sample Formal Report

Employee Lifestyle Study

Presented to
The Health and Safety Committee

Prepared by
Robbie Pinkney, Organizational Research Officer

September 20, 2006

Employee Lifestyle Study 2

TABLE OF CONTENTS

Table of Contents	2
List of Figures	2
Summary	3
Introduction	4
Literature Review	4
Method	5
Findings	6
Benefits and challenges	6
Suggestions	6
Exercise habits	7
Eating habits	8
Family time	8
Conclusions	10
Recommendations	10
References	11
Appendix 1	12

LIST OF FIGURES

Figure 1: Exercise
Figure 2: Eating Habits
Figure 3: Family Time

(continued)

SUMMARY

In response to rising sick days and use of stress counselling, the Health and Safety Committee at D&M Manufacturing initiated a study to help employees plan strategies in living healthy lifestyles. A focus group and a survey were employed to elicit suggestions and to discover current patterns of physical activity, nutrition, and the balance of work and family time. Some personnel seem to be doing well in leading a healthy lifestyle, but many are not, according to parameters established by Canadian health authorities. One-third reported that they do not participate in regular physical activity, two-thirds said they do not eat the required amount of fruit and vegetables, and almost two-thirds said they have little or no family fun time without planned activity or television. The recommendations, which reflect the suggestions made by employees, are to establish a lifestyle planning committee, improve food offerings in the cafeteria, install a gym, and allow a small amount of time during work hours for exercise.

Employee Lifestyle Study 4

INTRODUCTION

In March 2006, the Health and Safety Committee at D&M Manufacturing asked the organizational research officer to conduct a study that could help them understand how to assist employees in pursuing healthy lifestyles. The committee was responding to an apparent decline in health over the last ten years: the number of sick days rose by 20% between 1995 and 2005. In addition to this increase in sickness, our use of the employee assistance plan for stress counselling rose by 30% in the same period. The committee was also responding to management's expression of a desire to help. To help the committee in planning assistance, a literature review of current Canadian health guidelines was carried out, followed by a focus group and a survey.

LITERATURE REVIEW

According to the Public Health Agency of Canada (2004), a healthy lifestyle reduces stress, reduces illness, and helps family well-being. There is considerable variation in the way healthy people live but there are several commonalities, listed by the Agency as follows:

- Healthy lifestyles include a wide range of strategies, such as effective coping, lifelong learning, safety precautions, social interaction, volunteering, parenting, spirituality, balancing work and family, as well as good nutrition, physical activity, safe sex, and avoiding tobacco and substance abuse.
- A healthy lifestyle includes individual, community, and "interdependence between individual and community" dimensions.
- A healthy lifestyle is about striving to obtain a reasonable balance between enhancing one's personal health, the health and well-being of others, and the health of the community.

From the above list, the committee chose to focus on elements of lifestyle that could be affected by the company. After some discussion with management, the committee decided these would be physical activity, nutrition, and the balance of work and family time.

There is considerable research to show that physical activity is a good place to begin in living a healthier lifestyle. It is now understood that health is greatly improved by the fairly simple effort to engage in at least 30 minutes of

(continued)

moderate-to-vigorous exercise four times weekly (Canadian Fitness and Lifestyle Research Institute, 2005). The form of exercise can be nothing more vigorous than a brisk walk. Nevertheless, many Canadians do not currently achieve even this amount of exercise.

Nutrition habits also contribute greatly to health. The Public Health Agency of Canada (2004) states that people who eat at least 1200 millilitres (5 cups) of fresh fruits and vegetables are half as likely to contract colds and flu. A leading Canadian workplace and school nutrition expert has said that cafeterias must provide inexpensive salads, including dark leafy greens, and a variety of fresh fruits every day (Lalani, 2005). In addition, Lalani states that less healthy foods need to be priced higher than the healthy choices.

Finally, balancing family and work has now been recognized as a very important determinant of health. In a major Canadian study (Higgins, Duxbury, and Johnson, 2004), people who reported a high level of interference from work to family life were 2.4 times more likely to say their health is fair or poor than people with low interference. In addition, the study showed that people who experience high levels of caregiver strain make the greatest use of physicians' services.

Given the current knowledge of how these three lifestyle factors affect health, the committee decided to investigate our employees' needs in relation to them. They also wanted to establish a baseline of employees' habits so that they can measure how well new assistance efforts work.

METHOD

The study was divided into two parts between June and September 2006. First a focus group was held with ten employees from both the office and plant divisions. Participants were asked questions about how they felt their work life contributed to their well-being and how it might interfere with healthy living. Ideas on how the company might assist were elicited. The focus group session was audiotaped and transcribed for analysis, and its findings are included in this report.

The second stage of the study consisted of a survey sent to all 978 employees asking about their exercise habits, eating habits, and family time, and also seeking suggestions about how the company could help them. Note that 78% of our personnel our female. There was a very high response rate with 70% of questionnaires returned. The results were tabulated and are presented here.

FINDINGS

The focus group and survey yielded a rich body of information. Survey results supported the ideas that came out in the group discussion. The need for assistance is clear, and the suggestions that people made provide a good starting point for planning.

Benefits and challenges

Study participants said that they generally enjoyed their work and would miss the sense of accomplishment and the social interaction if they were not on the job. At the same time, they reported difficulties with finding time to prepare healthy food and to enjoy family-related activities. Their difficulties included the following:

- Fatigue and back problems from standing for long periods of time among plant staff
- Neck, wrist, and shoulder problems from computer time among office staff
- No time to make lunches
- Both spouses too tired to cook so fast food at dinner on many nights
- Weekends packed with yard work, kids' practices, housework and laundry
- Often unable as a result of overtime to go to children's planned activities
- Unused memberships at fitness clubs
- Constant fast pace of living

Suggestions

Participants were especially interested in healthier, fresher food choices in the cafeteria and in finding a way to exercise at work. Many said that they would take advantage of an exercise room with showers, and some suggested that they would be more likely to do this if 15 minutes could be added to the lunch hour, or to one break a day, for those who used the room. They also said that they would welcome healthier foods in the cafeteria as long as they were not too expensive and were truly fresh. In general, people want to continue working

(continued)

overtime hours, but they would like to be able to have their children's activities inserted into the overtime planning schedule to reduce conflicts.

Exercise habits

The amount of exercise that people reported varied considerably, but the majority of us don't get enough. The most active group (27%) said that they participate in the recommended minimum of 30 minutes of moderate to vigorous activity at least three times per week. More people (40%) said they did mild exercise three times a week, and a full third of our workers said they do none. See Figure 1.

Figure 1: Exercise

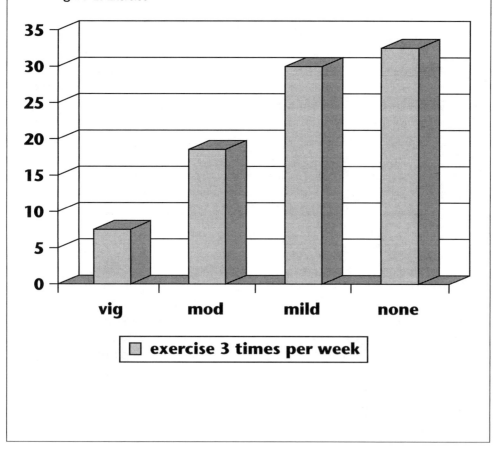

Eating habits

In terms of eating habits, very few people (5%) eat the maximum recommended amount of 1200 millilitres (5 cups) of fruit and vegetables per day. A larger group (28%) eats at 3–4 cups per day. That leaves just over two-thirds (68%) who do eat well below the required amount. See Figure 2.

Figure 2: Eating Habits

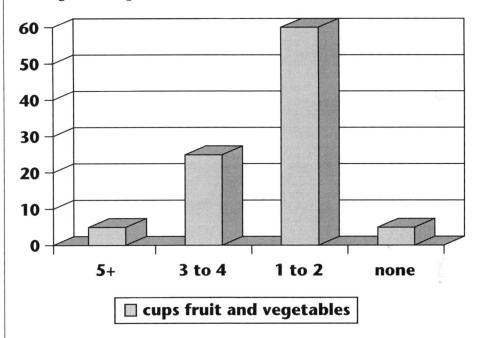

Family time

Balancing family life and work seems to be a challenge for many of us. Only 6% said they had family fun time 6 or more hours a week, without television or planned sports activities. A surprisingly large proportion (40%) reported only 1 to 2 hours a week, and 24% reported none. This means that almost two-thirds of our staff have very little "just plain fun time" with their children or spouses.

(continued)

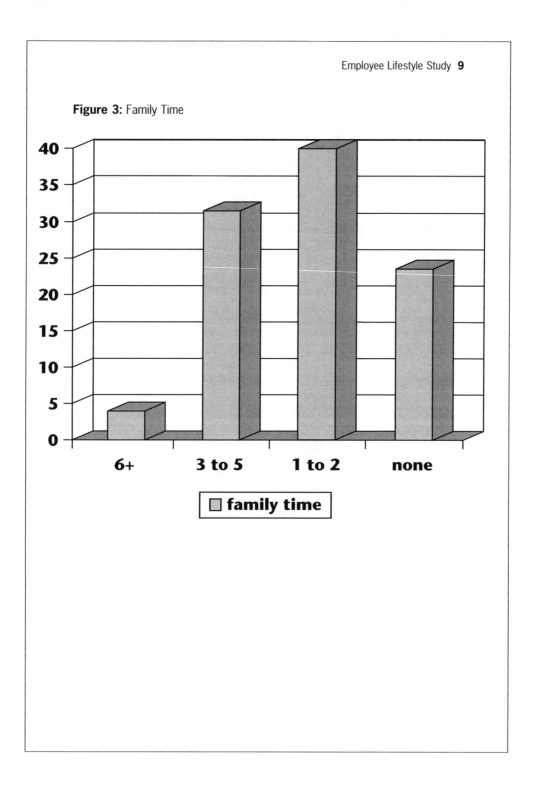

Figure 3: Family Time

Employee Lifestyle Study 10

CONCLUSIONS

The results indicate that there is considerable room for improvement in our lifestyles. A majority of us do not have enough physical activity in our lives, and many people would like to do more. Many of us do not consume enough fresh fruits and vegetables, and with busy lives, we find it difficult to prepare these foods for ourselves. The results also show that many of us have little or no family fun time that is free of planned activities or television. The challenges of exercise and food can be addressed by changes in cafeteria offerings and more opportunity for exercise at work. The issue of family time is more difficult to address, and further discussion is needed as to how the company might assist, beyond adopting the suggestion that children's regular activities be inserted into the overtime planning schedule. The company management has stated its commitment to assisting with efforts arising from the recommendations below.

RECOMMENDATIONS

1. Establish a lifestyle planning team to carry out the efforts outlined below and to plan and monitor future efforts. The committee would report to the Health and Safety Committee. Meetings during work time of two hours a week will be needed in the first month and two hours per month on an ongoing basis. The company vice-president is needed on the committee to help in planning and to ensure the company's contributions are carried out.
2. Evaluate cafeteria offerings and negotiate needed changes with the food provider.
3. Install a small gym to be planned by the new committee.
4. Initiate a policy of allowing 15 minutes added to one break or to the lunch hour for people who use the gym or go for walks (to be monitored via a simple exercise diary submitted to the new committee; form to be provided).
5. Initiate a procedure for submitting children's activity times to be placed in the overtime planning schedule.

(continued)

REFERENCES

Canadian Fitness and Lifestyle Research Institute. (2005). Canadians' physical activity: Assessing trends from 1998 to 2003. Retrieved April 6, 2005, from http://www.cflri.ca/pdf/e/2002pam.pdf

Higgins, C., Duxbury, L., & Johnson, K. (2004). Exploring the link between work-life conflict and demands on Canada's healthcare system. Public Health Agency of Canada. Retrieved April 6, 2005, from http://www.phac-aspc.gc.ca/publicat/work-travail/report3/pdfs/fvwklfrprt_e.pdf

Lalani, A. (2005). *Food is health*. Toronto: City Health Publications.

Public Health Agency of Canada. (2004). Healthy lifestyle: Strengthening the effectiveness of lifestyle approaches to improve health. Retrieved April 6, 2005, from http://www.phac-aspc.gc.ca/ph-sp/phdd/docs/healthy/chap1.html

Employee Lifestyle Study 12

APPENDIX 1

Employee Survey

The Health and Safety Committee is looking for ways to help our company's employees live healthy lifestyles. The results of this survey will help the committee understand your needs and plan appropriate assistance.

1. Please circle the statement that most closely describes your habits:

    I do mild exercise (slow walk, vacuuming) normally for at least half an hour at least 3 times a week.

    I do moderate exercise (brisk walk) normally for at least half an hour at least 3 times a week.

    I do vigorous exercise (athletic activity such as running, tennis, hockey) normally for at least half an hour at least 3 times a week.

    I do not exercise on a regular basis.

2. I eat the following amount of fruit and vegetables in a day:

    5+ cups   3–4 cups   1–2 cups   none

3. I have "fun time" with my family without planned activities or television:

    6+ hours/week   3–5 hours/week   1–2 hours/week   none

4. Please tell us what you feel the company could do to assist you in leading a healthy lifestyle.

3. GRAPHIC ELEMENTS OF REPORTS

Graphics are visual devices used to organize data and demonstrate relationships. Tables, graphs, photographs, and drawings are commonly used graphics. Graphics help your audience comprehend quantitative and structural information. Modern software programs make adding graphic elements easy, but it is important to know when and how to use these elements.

When and How to Use Graphics

As discussed in Chapter 8, the appropriate use of visuals can help to enhance your oral presentation. Similarly, in written reports, graphics add variety and interest, especially if the report is long. More important, they communicate well to visual thinkers. A graphic is appropriate if it saves a significant amount of reading time and increases understanding. When you are planning a report, look for sections that could be shortened and/or clarified by using a suitable graphic. Different types of graphics, such as lists, tables, and graphs, effectively convey different types of information. Fortunately, word processing programs make it easier to include graphics than in the past. By referring to Help, you can learn how to create graphics quickly.

If you decide to use a particular graphic, the next question is, how? No matter how important graphics are in a given report, they must be supported by text. This means that every graphic should be directly referred to in the report. With the exception of small informal lists or tables, graphics should be numbered and titled to make reference easier. The text should tell the reader the function of the graphic (for example, confirmation, analysis, or identification) and how to use it. If possible, keep graphics within the body of the text. If a graphic element needs a page to itself, keep it as close as possible to the textual reference. In any case, graphics should always come **after** the reference. Less important data may be collected at the end of the report in an appendix. All graphic elements, except simple lists, should be numbered and given a title, and they should be indicated in a list of figures placed after the table of contents.

Lists

The simplest form of graphic is the list. A list identifies items and places them in sequence. If the sequence follows a natural or logical order, such as chronological order or order of importance, the items should be numbered or lettered. If the order is not significant, the items can be left bare, or preceded by dashes or bullets. Lists are usually fully integrated into the text, without titles or identifying numbers, and they do not appear in the list of figures.

Tables

A table is an arrangement of data into columns (vertical) and lines (horizontal). If you compare the following paragraph with Table 1, you will see the value of using tables. Both the paragraph and the table present the same information.

As of April 30, the Acme Screw Company had 8913 employees. Of these, 2228 (or 25 percent) had less than five years' service; 1783 (or 20 percent) had five to ten years' service; 1426 (or 16 percent) had ten to fifteen years' service; 891 (or 10 percent) had fifteen to twenty years' service; 1070 (or 12 percent) had twenty to thirty years' service; and 1515 (or 17 percent) of the employees had thirty or more years' service with the Acme Screw Company.

Table 1: Acme Screw Employees' Length of Service

Years of Service	Number of Employees	% of Total
Under 5	2228	25
5 to 10	1783	20
10 to 15	1426	16
15 to 20	891	10
20 to 30	1070	12
30 or more	1515	17
Total as of April 30	8913	100 percent

Although the information given here is not very complex, the paragraph form makes it difficult for the reader to make sense of all the numbers. The table permits the reader to take in all the data at a glance, making comprehension and interpretation much easier. Tables are useful when you want to present characteristics of related objects, processes, and so on in a compact and precise way, to assist your reader to make comparisons and generalizations.

Tables have titles that identify the contents, and a number that directs the reader from the text to the table. The number and title are placed above the table. The source of the data is indicated in the title or in a footnote to the title. Always keep headings in parallel grammatical form. If you need to include a lengthy table in your report, place it in the appendices. Consider presenting the information in a graphic format, which will have much more impact.

Graphs

Graphs are used for mapping statistical information. For example, if your company has experienced an increase in sales over the last year, a bar graph showing sales for each quarter would be ideal. Since the purpose is to help the reader see the significance of data quickly, the impact of a graph should be immediate. To ensure this, make graphs simple and reasonably large. Use visual devices to help with their interpretation. Some of these will be discussed in reference to specific kinds of graphs. Like tables, graphs must have a number and a title. Graphs must always be mentioned in the report.

LINE GRAPHS

The line graph, which shows on a grid the relationship between variables, is often the best one to use for purely technical information. In a line graph, the behaviour of two variables is indicated. Generally, the **independent variable** is shown on the horizontal scale. Independent variables include measures of time, distance, and load. **Dependent variables,** those whose value is affected by changes in the independent variable, are represented on the vertical scale. Common dependent variables are temperature, rate, and cost. Generally, the reader uses a graph to get a picture of the relationship between two things, so clarity of overall impression is very important.

You can create a line graph with software quite easily. Variables and units of measure must be labelled. Then the data are plotted and the points are connected to make a continuous line. Graphs may represent the curve of either continuous or changing data. If the latter is represented, the lines should be clearly distinguishable. Figure 14.4 is an example of a line graph showing the changing temperature in Dartmouth, Nova Scotia.

BAR GRAPHS

Bar graphs represent relative quantities by using vertical or horizontal bars of varying lengths. Comparisons shown on a bar graph have more dramatic impact than those shown on a line graph. Again, it is fairly easy to create bar graphs using software. A bar graph usually has a scale, with the independent variable at the base of the bar, either at the bottom (in a vertical graph) or on the left side (in a horizontal graph). Bars are the same width and the same distance apart. Shading prevents confusion between bars and spaces. The bars on a horizontal graph are customarily arranged in order of decreasing or increasing length. In addition to using the scale, you may put exact numbers above or on the bar. Figure 14.5 gives you a good example of how bar graphs can help you present and distinguish data.

FIGURE 14.4 Sample Line Graph: Mean Average Temperature in Dartmouth, Nova Scotia

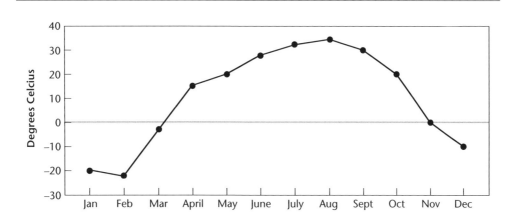

FIGURE 14.5 Sample Bar Graph: Sales Figures

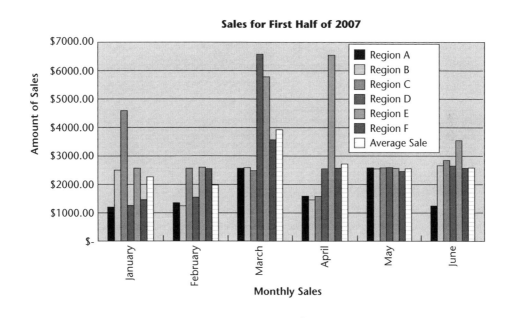

MAKE YOUR REPORT VISUALLY ATTRACTIVE

Since reading a report is always a chore, it is important to give it an inviting appearance and to give as much help as possible to the reader.

- Be alert to conventions in your organization or field, and follow them where conformity is expected.
- Use a letter-quality paper.
- Consider using colour, especially if the report contains graphics. However, do not overuse colour.
- Consider using a report template provided by your word processing software. You may choose to create your own template. If you will be issuing reports periodically, you will find a template very helpful, and your readers will appreciate a consistent look.
- Create a lean, uncluttered effect, using spacing and margins effectively.
- Consider using a range of typefaces to break up the "wall of print" effect.
- Give each level of heading a unique visual character. Establish a visual hierarchy with your most elaborate effects—centred and a different typeface, for example—reserved for major headings, and with the most modest choice, such as indented to lower case, for the smallest division of your text. Remember that using the heading levels provided by your software will allow you to automatically create a table of contents when the report is finished.

PIE CHARTS

A pie chart is a circle whose slices show a percentage distribution of the whole. Conventionally, slices are measured clockwise from the twelve o'clock position, beginning with the largest slice. Since it is difficult for readers to estimate the relative size of the slices, they must have labels and exact percentages on or beside them. Items that are less than 2 percent of the whole should be grouped as "miscellaneous" or "other," with the individual percentages given in parentheses or in a footnote. For an example of a pie chart, look at Figure 14.6.

FIGURE 14.6 Sample Pie Chart: Population of Betteridge, Ontario

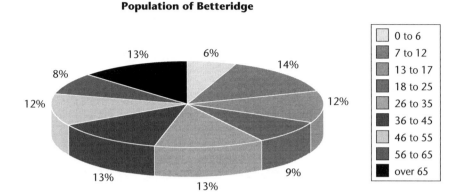

4 FINAL EDITING

Before a book, magazine, or newspaper is published, the contents are carefully checked by an editor. The editor checks facts, corrects mistakes, and tidies up style, but he or she also looks at the larger picture, the overall effectiveness of the piece of writing. Perhaps the order of some of the information should be changed. Perhaps there are redundancies or places where the writer wandered off topic. Something may have to be added to clarify a point or show its relevance to the main subject. Because an editor can stand back and see a piece of writing from a fresh, detached standpoint, the editor offers a great service to a writer. But at work, you will generally have to be your own editor. How can you do it? Here are some suggestions:

1. Put your first draft away for at least a day before editing it—longer, if possible.
2. Read it all the way through for an overview. If you see something that needs correcting or just doesn't sound or fit right, mark it, but don't stop to correct it.
3. Reread, looking for anything that you marked as out of place or irrelevant. Decide whether to move it, remove it, or reword it to make it fit the context.
4. Look at each sentence for wordy or unclear expressions. Watch for clichés and buzzwords, especially those that recur regularly. Rewrite sentences to remove these stylistic flaws.

5. Reread the whole report to see if your changes have maintained the flow and sense you intended.
6. Proofread your report for correct spelling, punctuation, and sentence structure. Get help with this if you know you have problems in any of these areas. Small mechanical errors can destroy the impact of your page. Using spell or grammar checkers is helpful, but remember that they will not pick up all errors.
7. If time permits, leave another day for a final reading before you have a final copy printed.

CHAPTER REVIEW/SELF-TEST

1. Word processing programs allow the writer to
   (a) incorporate professional-looking graphic elements easily.
   (b) prepare reports quickly without much planning.
   (c) rearrange and edit material to achieve optimal organization.
   (d) a and c.
   (e) all of the above.

2. Formal report organization may be
   (a) direct or indirect.
   (b) determined by the convention used in your organization or field.
   (c) created only according to the guidelines in this book.
   (d) a and b.
   (e) all of the above.

3. A transmittal document
   (a) will be in letter format if it is to go outside your organization.
   (b) will be in memo format if it is to stay inside your organization.
   (c) may be an e-mail message if the report is sent electronically.
   (d) may make a brief statement about the contents of the report.
   (e) a and b.
   (f) all of the above.

4. Which of the following statements is false?
   (a) A summary can be composed very quickly as it is seldom read by anyone.
   (b) A summary always appears on a separate page.
   (c) It may vary in its placement in the report.
   (d) It tells the "whole story in a nutshell."
   (e) It is always written last.

5. Which of the following statements is false?
   (a) A table is not as good as a paragraph for presenting series of numbers.
   (b) Tables should be listed in the list of figures.
   (c) Graphs present information with more impact than tables.
   (d) Long tables that are necessary to the report should be placed in the appendices.

6. A pie chart is used for showing
   (a) changes over time.
   (b) the parts of a whole.

7. Line and bar graphs are good for
   (a) showing changes over time.
   (b) providing strong visual impact.
   (c) helping the reader see the significance of the data quickly.
   (d) b and c.
   (e) all of the above.

8. Which of the following statements is false?
   (a) A report should follow the conventions in your organization or field.
   (b) Using a report template is helpful if you expect to write reports periodically.
   (c) People will think you are lazy if you use the same report format in successive reports.
   (d) A report should give a lean, uncluttered effect.

9. Which of the following statements is true?
   (a) It is all right if there are a few typos in the final report as long as they are not on the first page.
   (b) It is best to leave some time for the draft report to "cool off" before you do the final proofreading.
   (c) Spelling and grammar checkers will correct all your errors.
   (d) It is never all right to enlist a colleague's help in final proofreading.

10. Which of the following opinions do you share?
    (a) You get better at report writing when you do it often.
    (b) Report writing is a waste of time when you could be getting more important jobs done.
    (c) Handing over a finished report is very satisfying.
    (d) Writing high-quality reports will enhance your career prospects.

Answers

1.D 2.D 3.F 4.A 5.A 6.B 7.E 8.C 9.B 10. Defend your choice

EXERCISES

1. Read the guidelines for summaries on page 335 and introductions on page 336. When people start writing reports, they often confuse these two parts. Explain in your own words the similarities and differences between a summary and an introduction.

2. Writing summaries requires practice. Choose a long article from a magazine or trade journal. Write a single paragraph summary and submit it along with the article to your instructor. For practice in documentation, provide the full reference for the article using the documentation style preferred by your instructor.

3. Using the guidelines on page 332, write a transmittal memo to accompany the example formal report that begins on page 338. The memo should be addressed to the Health and Safety Committee.

4. Working in teams of three or four, take on some research needed for improving the exterior environment at your campus. Prepare a formal report of approximately 1200 words, giving your recommendations. Be sure to research the online literature that deals with such issues first. Then make your own observations about needed changes in such elements as leisure spaces, recreational areas, signage, lighting, trash disposal, walkways, and traffic patterns. Interview several college employees to get their views on what changes are needed. Finally, visit two other college campuses in order to make comparisons. Correctly cite the sources of your information and use the documentation style required by your instructor. Consider how your recommendations will be received by management (improvements will cost money) before deciding on direct or indirect order.

5. Research, plan, and write a report on a research question appropriate to your field. Ask your instructors and other experts for research questions for which they would like to know the answer. Your instructor may require a proposal and a progress report (see the Exercises in Chapter 12).

APPENDIX

Handbook of Grammar and Usage

INTRODUCTION

In the following pages, you'll find a short summary of the most important principles of grammar, punctuation, spelling, and usage.

This appendix does not pretend to be a complete and thorough guide to all the intricacies of English grammar; rather, it offers you brief, streamlined advice on how to handle those troublesome areas of writing that give college students the most difficulty.

In putting this appendix together, we've tried to stick to the basic information that you'll really need to write at college and on the job. If you need a more complete guide to grammar, you should consult a full-length English handbook, such as *Harbrace Handbook for Canadians*, Sixth Canadian Edition, by John C. Hodges and Andrew Stubbs (Toronto: Thomson Nelson, 2003).

To some students, the word *grammar* conjures up images of frowning teachers who rap children over the knuckles for every conceivable infraction of the rules. The fact is, nonetheless, that grammar, rather than being a maze of useless rules, is actually a necessity if we are to communicate efficiently. It provides the logical, mutually recognizable patterns in which language operates.

RECOGNIZABLE PATTERNS

Consider the following two sequences of numbers:

(a) 5 3 9 1 4 0 8 2 7 6
(b) 9 8 7 6 5 4 3 2 1 0

Which sequence puts less of a burden on the reader? Which sequence is easier to remember? Why? Sequence **b** is the answer, of course, because it follows a recognizable pattern. We look for such patterns in every aspect of life — in our daily schedule, in the professors' lectures, in the music we listen to, in the way we drive our cars, in social protocol, and in the grammar of our language.

Take a look at the following two sentences:

(c) audited accountant ledger The the.
(d) The accountant audited the ledger.

Which of these sentences communicates more efficiently? Which is easier to remember? Why? The answer is **d**, because it follows the grammar (recognizable patterns) of the English language, whereas **c** does not.

UNITS IN GRAMMAR

"Give me a place to stand," said Archimedes, "and I will move the world." By analogy, in grammar we could say, "If you can write a sentence, you can move the world." The chief objective of this Appendix is to enable you to write a sentence. To accomplish this objective, we'll look at four elements of an English sentence:

1. **Words** — the basic element of all communication
2. **Phrases** — groups of words that function as single words
3. **Clauses** — groups of words with a subject and a verb
4. **Sentences** — clauses that express a complete idea

Words

Words serve several functions. Here are some of the most common:

1. **Nouns** name persons, places, things, or qualities. They generally form the plural by adding an *-s* to the singular (*girl→girls*), but a few are irregular (*datum→data, series→series, woman→women*).
2. **Pronouns** are stand-ins for nouns or noun phrases. Their forms often show whether they are functioning as subjects, objects, or possessives and whether they refer back to singular or plural nouns. Table A.1 summarizes the forms of the three most common kinds of pronouns: personal pronouns, which refer to people; relative pronouns, which introduce adjective clauses and "relate"

TABLE A.1 Common English Pronouns

Pronoun Type	Person	Singular			Plural		
		Subject	**Object**	**Possessive**	**Subject**	**Object**	**Possessive**
Personal	1st	I	me	my/mine	we	us	our/ours
	2nd	you	you	you/yours	you	you	you/yours
	3rd	he	him	his	they	them	their/theirs
		she	her	her/hers			
		it	it	its			
Relative	3rd	who which that	whom	whose	who which that	whom	whose
Interrogative	3rd	who? what?	whom?	whose?	who? what?	whom?	whose?

them back to the noun being described; and interrogative pronouns, which ask questions.

3. **Verbs** express action or states of being. Each verb has three parts: (1) the stem (*audit/write*), (2) the past (*audited/wrote*), and (3) the past participle (*audited/written*). Verbs have two basic tenses — present (*audit*) and past (*audited*). With auxiliaries, verbs can also express future time (*is going to audit, will audit*) as well as shades of present time (*is auditing*) and past time (*has audited, had audited*). Most verbs have a passive form (*is audited*), which differs from the active in that the passive contains (1) some form of *be/am/is/are/was/were/been/being* plus a past participle and (2) directs its action toward the grammatical subject (*the ledger was audited*).

4. **Modifiers**
   (a) **Adjectives** modify or clarify nouns (I audited *the red* ledger; *the* ledger that I audited was *red*). Two classes of adjectives—the demonstratives (*this/that/these/those*) and the articles (*a/an/the*)—must always precede the noun; they also differ from other adjectives in having to agree with the noun in number (*these book* is not grammatical English, and neither is *a books*).
   (b) **Adverbs** modify verbs. Most end in *-ly* (I wrote *quickly*), but some don't (I wrote *fast*). Occasionally, adverbs team up so that one modifies the other (I wrote *unusually fast*).
   (c) **Intensifiers** generally emphasize the meaning of adjectives, participles, or adverbs (the case history was *very* interesting). *Very*, incidentally, is not an adverb because *very* cannot modify a verb. For example, we can say *the bookkeeper was very deceitful*, but not *the bookkeeper deceived very*.

5. **Prepositions** are words such as *at, by, into, under, above,* and *from*, which introduce prepositional phrases (a preposition plus a noun or pronoun) and connect the phrase to the rest of the sentence (I went *into my office* to audit the ledger).

6. **Conjunctions** join two or more grammatical units. Coordinating conjunctions join two or more units that are grammatically equivalent (two prepositional phrases, two verbs, two nouns, etc.). The coordinating conjunctions are *and, or, nor,* and *but*. When used to combine two independent clauses, the coordinating conjunction should be preceded by a comma (the ledger contained mistakes, *and* the accountant had to work overtime to find them).

Subordinating conjunctions introduce dependent clauses (for a definition, see the section on clauses) and connect them to the rest of the sentence, showing the precise relationship between two ideas. Some common subordinating conjunctions are *because, if, unless, while, when, after, so that*, and *although*. When a subordinating conjunction comes at the beginning of the sentence, the clause it introduces is followed by a comma (*when you send us a cheque*, we will send you the books). However, the comma is not necessary when the dependent clause is at the end of the sentence (we will send you the books *when you send us a cheque*).

7. **Expletives.** The two common grammatical expletives are *it* and *there*. Both serve merely to get a clause moving (*it is raining*), but *there* can never be the subject (say *there ARE two ledgers on the desk* and *there IS one ledger on the table*).

8. **Other Words.** In the "other" category, English has one significant group — the interjections. These include *oh* and *well* used at the beginning of a clause.

Phrases

Phrases are groups of words that work together grammatically; they function as a single word would function if one were available. Common types of phrases include:

1. **Noun phrases,** as in *the accountant*, which always function as nouns.
2. **Verb phrases,** as in *is auditing*, which always function as verbs.
3. **Infinitive phrases,** as in *to audit*, which may function as nouns (Laurie decided *to audit* the ledger), as adjectives (the accountant *to audit* the ledger is Laurie), or as adverbs (*to audit* the ledger, you will need to see the bursar).
4. **Participial phrases,** which always function as adjectives — modifiers of nouns or pronouns (*auditing the ledger*, the accountant had difficulty with some entries; *audited according to the specifications*, the ledger was returned to the file). The present participle always ends in *-ing*, and the past participle of most verbs ends in *-ed*.
5. **Gerundive phrases,** which always contain an *ing* word formed from a verb and used as noun — unlike the present participle, which is used as an adjective (*working late* is unavoidable for most accountants during the last two weeks of April).
6. **Prepositional phrases,** which may function as either adverbs (the accountant testified *on the witness stand*) or adjectives (the accountant *on the witness stand* is Laurie Stanfield).

Clauses

A clause must contain two things:

1. **A subject,** which is the doer of the action with an active verb and the receiver of the action with a passive verb. In commands, the subject is often understood ([you] type this letter).
2. **A verb,** that is, a full verb such as *audit*, not an infinitive such as *to audit*.

Clauses may be either independent or dependent.

1. **Independent clauses,** which are sometimes called main clauses, express a complete idea and can stand alone as sentences (*the accountant audited the ledger*).
2. **Dependent clauses,** which are sometimes called subordinate clauses, express an unfinished idea and cannot stand alone (*when the accountant audited the ledger*). A dependent clause may function as an adverb (*When the accountant audited the ledger*, the judge dismissed the lawsuit), as an adjective (The accountant *who audited the ledger* is Laurie Stanfield), or as a noun (The judge could see *that the accountant had audited the ledger*).

Sentences

A sentence may be of any length provided that it

1. begins with a capital letter,
2. contains at least one independent clause,
3. ends with a period or other appropriate terminal punctuation mark.

Sentences may be simple (one independent clause), compound (two or more independent clauses joined by a coordinating conjunction), or complex (an independent clause plus one or more dependent clauses).

Quiz

This quiz may help you determine how well you understand the principles just presented. Use the following list to designate the problem in each sentence:

A. Error in agreement or usage

B. "Sentence(s)" with no independent clause

C. Correct, no error

D. Erroneous statement about grammar

E. Error in punctuation

_____ 1. When he had written his report and handed it to his boss.
_____ 2. The passive voice in grammar is a combination of some form of be plus a past participle.
_____ 3. Students often worry about jobs. Where they can find them and how they can get them.
_____ 4. What he said when he left on holidays.
_____ 5. Please be on time for your appointment.
_____ 6. The financial planner applied the new formula and the long-term forecast became more optimistic.
_____ 7. A worker must pay their union dues by payroll deduction.
_____ 8. The accommodations were inadequate.
_____ 9. Now that you have finished the quiz, is there any questions?

Here are the answers and corrections for the quiz:

1. _B_ When he had written his report, he handed it to his boss.
2. _C_
3. _B_ Students often worry about jobs. Where they can find jobs and how they can get them are primary concerns. (Or: Students often worry about where they can find and how they can get them.)
4. _B_ When he left on holidays, he said he would return in three weeks.
5. _C_
6. _E_ The financial planner applied the new formula, and the long-term forecast became more optimistic.
7. _A_ Workers must pay their union dues by payroll deduction.
8. _C_
9. _A_ Now that you have finished the quiz, are there any questions?

GUIDELINES FOR USAGE

Having laid the framework of the recognizable patterns of English, we are ready for more detailed guidelines on ways to reinforce these patterns. The guidelines, if followed, will enhance your ability to communicate, especially in writing, so that the reader (or listener) can understand you more easily.

In addition to the notations used in the guidelines, the symbols in Figure A.1 are often used in proofreading to correct various errors.

FIGURE A.1   Standard Proofreading Marks

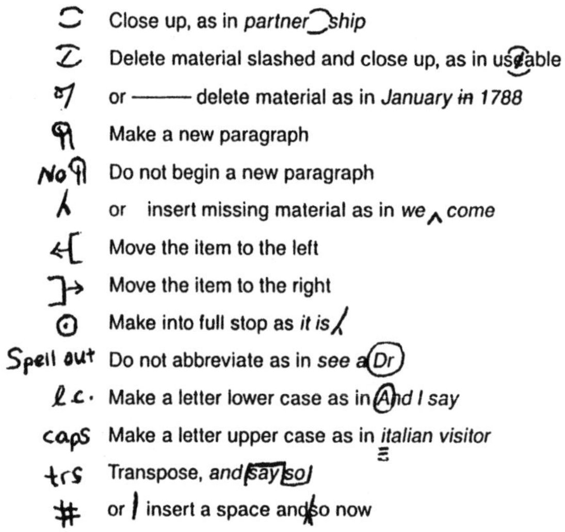

⌒   Close up, as in *partner⌒ship*
⌫   Delete material slashed and close up, as in *us/able*
⌿   or ——— delete material as in *January ~~in~~ 1788*
¶   Make a new paragraph
No ¶   Do not begin a new paragraph
∧   or   insert missing material as in *we ∧ come*
⊣[Move the item to the left
]⊢   Move the item to the right
⊙   Make into full stop as *it is/*
Spell out   Do not abbreviate as in *see a ⓓⓡ*
l.c.   Make a letter lower case as in *Ⓐnd I say*
caps   Make a letter upper case as in *italian visitor*
trs   Transpose, *and say so*
or / insert a space and so now

Confusion of "A/An"

Use *a* before consonants, *an* before vowels.

COLLOQUIAL
The company requested a audit.

STANDARD
The company requested an audit.

Confusion of Adjective/Adverb

Do not misuse an adjective for an adverb or an adverb for an adjective.

MISLEADING
Laurie felt badly about the erroneous entries in the ledger.

STANDARD
Laurie felt bad about the erroneous entries in the ledger.

COLLOQUIAL
You did real good.

STANDARD
You did really well.

COLLOQUIAL
I could sure use a new calculator.

STANDARD
I could surely use a new calculator.

Some of the pairs cited in "Word Misuse" (page 377) are adjective/adverb confusions (for example, *because of/due to* and *all together/altogether*).

Adverbial Noun

Do not use an adverb clause where a noun clause is needed. The words *if* and *because* are adverbs and should not be used to introduce noun clauses. *When* and *where* are also subject to this error when they are used in definitions.

INSTEAD OF
I'll see if she is here.

USE
I'll see whether she is here.

INSTEAD OF
The reason is because I'm penniless.

USE
The reason is that I'm penniless.

INSTEAD OF
Metathesis is when two letters are transposed.

USE
Metathesis is the transposition of two letters.

OR
Metathesis occurs when two letters are transposed.

Dangling Expressions
(see also "Misplaced Modifiers," page 369)

Make sure that a participle has a noun or pronoun to modify logically. Participles are verbal forms ending in -*ing* (present) and -*ed* (past) that are used as adjectives. If the noun or pronoun that the participle is supposed to modify is not actually named, the participle is said to dangle.

DANGLING PARTICIPLE
Rejected by the management, a union strike began at midnight.

CORRECT
Its offer rejected by the management, the union went on strike at midnight.

The most vexing problem occurs with a present participle at the beginning of a sentence (often a passive sentence). Here is how such a dangling participle develops.

THOUGHT 1
I attended the class.

THOUGHT 2
I used a cadaver for experiments.

THOUGHT 1A
Attending the class (*present participle*)

THOUGHT 2A
A cadaver was used for experiments. (Note that the passive form of the sentence, in eliminating the need to mention the doer of the action, simultaneously eliminates the word that *Attending* needs to modify.)

DANGLING PARTICIPLE
Attending the class, a cadaver was used for experiments.

CORRECT
Attending the class, I used a cadaver for experiments.

CORRECT
In one of my classes, a cadaver was used for experiments.

Be sure that a gerund does not dangle and thus imply an incorrect subject.

DANGLING GERUND
By standing on the riverbank, a steamboat could be seen.

CORRECT
By standing on a riverbank, I could see a steamboat.

Incomplete Constructions

Do not leave out any word that is necessary to make a statement or a comparison logical and complete.

INCOMPLETE
Richard wanted to pass not only the LLB examinations but the bar examination.

COMPLETE
Richard wanted to pass not only the LLB examinations but also the bar examination.

INCOMPLETE
Be specific as possible.

COMPLETE
Be as specific as possible.

Misplaced Modifiers
(see also "Dangling Expressions," page 368)

Place a modifier and the word it modifies as close together as possible.

MISPLACED
It is unwise to carry an electromagnet into a computer centre that is activated.

BETTER
It is unwise to carry an electromagnet that is activated into a computer centre.

BETTER STILL
It is unwise to carry an activated electromagnet into a computer centre.

   A particular problem arises with the use of words such as *almost*, *just*, *nearly*, and *only*. Observe how *only* affects meaning when it is in each of the positions noted by the caret (∧):

$\overset{1}{\wedge}$ The $\overset{2}{\wedge}$ house $\overset{3}{\wedge}$ costs $\overset{4}{\wedge}$ $290 000 $\overset{5}{\wedge}$.

1. You'll pay an additional fortune for the garage!
2. There's no other house on the block.
3. The house really isn't worth that much.
4. The house doesn't cost any more than that.
5. There's no additional charge in pesos or yen.

Parallelism, Faulty

To make concepts that are parallel in thought parallel in structure, use the same grammatical form for all items in a series. In the following example, use either the present participle (*keypunching/writing*) or the infinitive (*to keypunch/to write*) for both items in the series:

FAULTY
Keypunching the cards and to write the program might take all week.

PARALLEL
To keypunch the cards and to write the program might take all week.

PARALLEL
Keypunching the cards and writing the program might take all week.

Be sure to follow each member of correlative pairs (*both ... and, either ... or, neither ... nor, not only ... but also*) with the same grammatical structure:

FAULTY
Both in cost and space, this computer is best.

PARALLEL
In both cost and space, this computer is best.

PARALLEL
Both in cost and in space, this computer is best.

Passive, Needless
(see also "Dangling Expressions," page 368)

Avoid needless passives when actives will do the job. (Don't forget, however, that a passive's lack of specificity can be useful at times—for example, when stating a refusal.)

PASSIVE (AND STILTED)
The audit was conducted by Ernst and Whinney.

ACTIVE (AND DYNAMIC, SPECIFIC)
Ernst and Whinney conducted the audit.

Preposition at End of Sentence

Since prepositions at the end of a sentence are native to the English language, avoid them only if by doing so you make the sentence more effective.

FLAT
This is the desk our best accountant worked herself to death at.

MORE EFFECTIVE
This is the desk at which our best accountant worked herself to death.

Usually, however, you can just forget about this rule.

Pronoun Problems

Give each pronoun a specific noun (antecedent) to refer to. Problems occur most often with *it*, *that*, *this*, *they*, and *which*.

VAGUE
The accounts had not been audited. This brought about many problems with Revenue Canada.

CLEAR
The accounts had not been audited. This failure brought about many difficulties with Revenue Canada.

CLEAR
Failure to audit the accounts brought about many difficulties with Revenue Canada.

Make each pronoun agree with its antecedent in number (see also "Subject–Verb Disagreement," page 374) and in gender.

1. Use a plural pronoun to refer to a plural noun and a singular pronoun to refer to a singular noun.
2. If the antecedent is two or more nouns or pronouns joined by *and* (as in *X and Y*), make the pronoun plural (such as *they*), unless *X* and *Y* form a single unit, as in *bacon and eggs* (*it*).
3. If the antecedent is two or more nouns or pronouns joined by *or* or *nor*, make the pronoun agree with the nearest antecedent.

4. If the antecedent is a collective noun, be careful. Use a singular pronoun if the collective antecedent is unitary (a single unit). Use a plural pronoun if the collective antecedent is individual (a collection of separate persons or things).

   UNITARY
   Management renegotiates its contract with the union every two years.

   INDIVIDUAL
   The management have agreed not to raise their salaries this year.

   Sometimes, however, the use of a plural pronoun with an individual collective antecedent is so awkward that it is better to rewrite the sentence.

5. Avoid sexual bias in using pronouns. Since some people have strong feelings on this matter, you are probably best advised to avoid the controversy by being sensitive to it.

   ✗ POOR
   A student should pay his fees.

   ✓ IMPROVED
   Students should pay their fees.

   ✗ POOR
   Call the operator and ask for her.

   ✓ IMPROVED
   Call the operator and ask.

   ✗ POOR
   The employee gradually becomes more concerned about his retirement benefits.

   ✓ IMPROVED
   The employee gradually becomes more concerned about the retirement benefits.

6. Treat words following *every* (everybody, everyone), *any* (anybody, anyone), and *no* (nobody, no one) as singular (*every man, woman, and child has his or her problems*).

7. Understand foreign plurals (*data, memoranda*) as plurals in English too.

Use the proper case of pronoun.

1. Remember that a pronoun takes the case (subject, object, possessive) that it has in its own clause.

✗ POOR
Give the file to whomever asks for it.

✓ IMPROVED
Give the file to whoever asks for it. (*Whoever* is the subject of *asks*: the object of the preposition *to* is the entire clause *whoever asks for it*.)

✗ POOR
This is her you're speaking with.

✓ IMPROVED
This is she you're speaking with. (The object of *with* is an "understood" *whom* or *that*.)

2. In formal contexts, use subject pronouns for all subject uses.

Susan and I wrote the program.
This is he, speaking.

3. Use object pronouns for all object uses.

The secretaries gave a party for John and me.
Between you and me there is an understanding.

4. Use the possessive case before a gerund.

✗ POOR
I was displeased with him resigning so abruptly.

✓ IMPROVED
I was displeased with his resigning so abruptly.

5. Be conscious of courtesy in sequence of pronoun; if you can put yourself last, do so.

Janice and I are the top salespersons.

Sentence Fragment

Give each sentence a subject and a verb. Remember that any initial subordinating word will prevent the phrase from forming a complete sentence.

FRAGMENT
Because the bookkeeper was drunk.

SENTENCE
The bookkeeper was drunk.

Often a sentence fragment is actually a clause or phrase that belongs with the preceding sentence but has erroneously been punctuated as a sentence by itself.

✗ POOR
Ryerson Polytechnic University is a technological university. The only one in Toronto.

✓ IMPROVED
Ryerson Polytechnic University is a technological university — the only one in Toronto.

✓ IMPROVED
Ryerson Polytechnic University is the only technological university in Toronto.

Split Infinitive

A split infinitive is usually awkward. Try to avoid it.

✗ POOR
Virginia decided to, at the last minute, take the RIA review course.

✓ IMPROVED
At the last minute, Virginia decided to take the RIA review course.

Subject–Verb Disagreement

Make each subject and verb agree in number.

1. Use a singular verb with a singular subject, a plural verb with a plural subject.
2. If the subject is two or more nouns or pronouns joined by *and*, as in *X and Y*, make the verb plural, unless *X and Y* forms a single unit (*ham and eggs is a common breakfast*).

✗ POOR
Toronto and Ottawa is the largest cities in Ontario.

✓ IMPROVED
Toronto and Ottawa are the largest cities in Ontario.

3. If the subject is two or more nouns or pronouns joined by *or* or *nor*, make the verb agree with the nearest substantive. (For verbs, this rule applies to agreement in person as well as to agreement in number.)

   POOR
   Either the accountants or the manager have objected to the new policy.

   IMPROVED
   Either the manager or the accountants have objected to the new policy.

4. Use a singular verb if a collective-noun subject is unitary. If a collective-noun subject is acting as individuals and a plural verb is awkward, it is better to rewrite the sentence.

   UNITARY
   The committee has remained firm in its resolve.
   The company has decided to expand. It will build two new stores.

   INDIVIDUAL
   The committee have disagreed on three items.

   BETTER
   The committee members have disagreed on three items.

   Absolute rules are difficult to make here, however, especially for Canadians. Some follow the American style of insisting that collective nouns, such as *group* and *committee*, take singulars; others use the British approach and accept these words as plurals. The same variations occur with company names. Some well-educated Canadian would write, "I have called Sears, and they replied ... ," while others would put it "it replied." Many would, however, use "the Bay ... it" because that company name does not end with an *s*.

5. In sentences beginning with *there* or *here*, be careful to make the verb agree with the logical subject.

   POOR
   There is too few accounting professors.

   IMPROVED
   There are too few accounting professors.

6. Treat words after *every* (*everybody, everyone*), *any* (*anybody, anyone*), and *no* (*nobody, no one*) as singular (*every man, woman, and child was present*).

7. Understand foreign-language plurals (*data, memoranda*) as plurals in English too.

✗ POOR
This data is the latest we have received.

✓ IMPROVED
These data are the latest we have received.

8. Make the verb agree with the real subject, not with the object of an intervening prepositional phrase.

✗ POOR
Yesterday's balance of the accounts were correct.

✓ IMPROVED
Yesterday's balance of the accounts was correct.

Tense Problems

1. Use English tenses properly with regard to time.
   (a) The present tense describes current happenings (I am studying accounting), facts that are always true (only women can give birth), or historical events discussed in present time (Hildebrand goes to Canossa and begs for mercy).
   (b) The past tense describes events in past time.
   (c) English expresses future time by using the present tense
      (1) with an adverb of time (I *study* accounting *tomorrow*);
      (2) with *will* or *shall* (I *will study* accounting);
      (3) through other means, usually the present participle of go (*I am going to study* accounting).

2. Use tense consistently.

✗ POOR
Geraldine adds up the columns and advised her supervisor about the overruns.

✓ IMPROVED
Geraldine added [or adds] up the columns and advised [or advises] her supervisor about the overruns.

3. If events happen at different times, use auxiliary verbs logically.

✗ POOR
Peri was a good skier before she has broken her leg.

✓ IMPROVED
Peri was a good skier before she broke her leg.

Word Misuse

1. Use logical comparisons. Some adjectives and adverbs are absolute in meaning and do not logically submit to comparison. Examples are *complete*, *full*, *perfect*, and *unique*. Instead of saying *fuller/fullest* or *more/most unique*, use *more/most nearly full* or *unique*.
2. Avoid needless use of ink (verbiage).

✗ POOR
Due to the fact that ...

✓ IMPROVED
Because ...

✗ POOR
Fill the tank up.

✓ IMPROVED
Fill the tank.

✗ POOR
Utilize

✓ IMPROVED
Use

✗ POOR
The ledger which was returned to me was Tom's.

✓ IMPROVED
The ledger returned to me was Tom's.

✗ POOR
General consensus of opinion

✓ IMPROVED
Consensus

✗ POOR
And etc.

✓ IMPROVED
And so on

3. Use standard expressions.
   (a) There is no -s on anywhere, nowhere, or a long way.
   (b) The adverbial *kind of* and *sort of* (or kinda, sorta) should be omitted or changed to a standard expression, such as rather, somewhat, or a little.
   (c) The infinitive sign *to* is preferable to *and* after try; for example, try to come, not try and come.
   (d) Some words in colloquial speech are out of place in formal usage:

Colloquial	Correct
enthused	enthusiastic
irregardless	regardless *or* irrespective
yourn, yous	yours, you
illiterate *past tenses and past participles, such as* brung, clumb, knowed, have went, *and* have wrote	correct *past tenses and past participles, such as* brought, climbed, knew, have gone, *and* have written

   (e) Some words are more specific than others:

Inexact	Exact
contact	communicate with, telephone, visit
great	famous, large, wonderful
nice	attractive, congenial, easygoing, thoughtful

(f) Some expressions that seem alike are actually different. Learn to discriminate between the following:

Accept	to receive gladly
Except	with the exclusion of (preposition); to leave out (verb)
Affect	to influence (verb)
Effect	result (noun) and to bring about (verb)
All ready	prepared
Already	before or previously
All together	in a group
Altogether	entirely
All right	completely right
Alright	a common misspelling of *all right*
Almost	nearly (adverb)
Most	greatest in number (adjective)
Among	refers to three or more
Between	refers to two
Amount	refers to things that cannot be counted
Number	refers to things that can be counted
Any one	refers to any person or thing of a specific group
Anyone	refers to any person in general
As … as (with)	correct in positive and negative comparisons
So … as (with)	correct in negative comparisons only
Because of	on account of
Due to	attributable to (used following the verb *to be*)
Can	refers to ability
May	refers to permission
Cite	to quote (verb)
Sight	the act of seeing (noun) or to see within one's field of vision (verb)
Site	a place (noun), such as *the Expo 86 site*
Continual	repeated regularly
Continuous	without stopping
Each other	refers to two
One another	refers to more than two
Farther	refers to literal distance
Further	refers to distance in time, degree, or quantity
Fewer	refers to number (*fewer accidents*)
Less	refers to quantity (*less money*)

Imply	to hint at
Infer	to draw a conclusion
In	indicates position or location
Into	indicates movement or direction
Its	possessive pronoun
It's	contraction for *it is*
Lay	to put, place, or prepare
Lie	to recline or be situated
Lead	to guide or to serve
Led	past tense and past participle of *lead*
May be	indicates possibility (verb)
Maybe	perhaps (adverb)
Oral	spoken rather than written
Verbal	associated with words, both spoken and written
Passed	past tense of *pass*
Past	no longer current (adjective)
Principal	first in importance (adjective); one who holds a primary position (noun); in finance, capital as distinguished from the interest on, or gain or loss from, that sum (noun)
Principle	a basic truth (noun)
Raise	to elevate
Rise	to move upward
Set	to put in position
Sit	to be seated
Stationary	not moving
Stationery	writing materials
Their	possessive form of *they*
They're	contraction of *they are*
There	in that place (adverb)
To	toward (preposition)
Too	in addition (adverb)
Two	number
Whereas	inasmuch as
While	refers to time
Who's	contraction of *who is*
Whose	possessive form of *who*
You're	contraction of *you are*
Your	possessive form of *you*

Quiz on Usage Guidelines

This quiz may help you to determine how well you understand the principles presented above. Correct all errors in usage that you find in the following sentences.

1. I sure appreciate your help on this project.
2. Suspecting fraud, the books were audited last month.
3. John plans to work to put himself through college and for experience in his profession.
4. After the accident, Jennifer insisted that she was alright.
5. Always use good quality stationary when you write letters of application.
6. I will not attend the meeting due to the fact that I have a report deadline tomorrow morning.
7. The chairman called the meeting to order.
8. I'll see if she wants to study with us.
9. Unemployment is high right now which is disturbing.

Here are the corrected sentences:

1. I appreciate your help on this project (very much).
2. Suspecting fraud, the judge ordered an audit of the books last month.
3. John plans to work to put himself through college and to gain experience in his profession.
4. After the accident, Jennifer insisted that she was all right.
5. Always use good stationery when you write a letter of application.
6. I will not attend the meeting because I have a report due tomorrow morning.
7. The chairperson called the meeting to order.
8. I'll see whether she wants to study with us.
9. Current high employment rates are disturbing.

GUIDELINES FOR PUNCTUATION

Apostrophe

1. Use the apostrophe to indicate the possessive case of all nouns. If the noun ends in *s*, add only the apostrophe; if the noun does not end in *s*, add *s* for the singular possessive. End the word with *s* in the case of the plural possessive.

Singular	Singular Possessive	Plural	Plural Possessive
Thomas	Thomas'	Thomases	Thomases'
Jane	Jane's	Janes	Janes'
company	company's	companies	companies'
woman	woman's	women	women's

2. Use the apostrophe only with the last noun in a series citing joint ownership.

   INDIVIDUAL OWNERSHIP
   John's and Mary's clothes.

   JOINT OWNERSHIP
   Derrill and Suzanne's advertising agency.

3. Use the apostrophe to indicate the possessive case of indefinite pronouns (*everybody's, one's*).
4. Use the apostrophe to stand for the missing elements in contractions (*doesn't, don't, we'll*).
5. Do not use needless apostrophes. The most flagrant violation of this rule is the confusion of contraction *it's* (for *it is*) with possessive pronoun *its*. Although some educated people use apostrophes in simple plurals of letters or numbers, there seems to be little justification for the practice. The following procedure is acceptable for indicating simple plurals: *1920s* or *CAs*.

Brackets

1. Use brackets to insert material into a quotation, as in the following example:

   Mr. Fogbound claimed that "unemployment is no longer an anecdote [*sic* antidote] to inflation."

2. Except in mathematics and in computer languages, use brackets for parentheses inside of parentheses. For example:

   According to Mark Lester, "Grammar is a way of talking about how words are used to make units that communicate a meaning" (*Introductory Transformational Grammar of English*, 2nd ed. [New York: Holt, Rinehart and Winston, 1976], p. 13).

Colon

1. Use a colon to introduce an explanation or a long quotation (particularly one that contains commas).

2. Use a colon to introduce a list:
   (a) When the list appears as a list on the page (as this one does)
   (b) When the list is added to a complete sentence regardless of the list's appearance on the page

   The colon is unnecessary when the list is part of the sentence. For example:

   UNNECESSARY COLON
   We sent technicians to: Toronto, Winnipeg, and Montreal.

   CORRECT
   We sent technicians to Toronto, Winnipeg, and Montreal.

   ALSO CORRECT
   We sent technicians to three centres: Toronto, Winnipeg, and Montreal.

3. Use a colon in the following particular places:
   (a) Between title and subtitle, as in Grinder and Elgin's *Guide to Transformational Grammar: History, Theory, Practice*
   (b) Between place of publication and publisher in a citation, as in *Toronto: Thomson Nelson, 2003*
   (c) Between hours and minutes, as in *11:05 AM*

Comma

1. Use a comma before a coordinating conjunction (*and, or, nor, but*) joining two independent clauses. For example:

   The financial planner applied the new formula, and the long-term forecast became much more optimistic.

2. Use a comma after an introductory word, phrase, or clause. For example:

   Yes, I plan to go on Saturday.
   In preparing this form, you should write everything in ink.
   When Bonnie finished the examination, she forgot to hand in the answer sheet.

3. Use a comma before a short direct quotation (but not before an indirect one). For example:

   ✗ POOR
   Eleanor said, that she would not go.

✓ IMPROVED
Jim said, "I'll be in the office all day."

4. Use commas to separate words in a series. For example:

   Mark wrote the COBOL program quickly, neatly, and accurately.
   Gretchen became involved in a long, expensive lawsuit.

5. Use commas to set off a relative clause that adds nonessential information to the sentence and could be left out without changing the meaning. For example:

   NONESSENTIAL RELATIVE CLAUSE
   Thomas Edward James, who sits in the back row, gave an excellent presentation.

   ESSENTIAL RELATIVE CLAUSE
   The tall man who sits in the back row gave an excellent presentation.

6. Use commas to set off qualifying or explanatory material. For example:

   Kimberly Shipman, our last supervisor, transferred to Nova Scotia.
   Take the blue form, not the red one, to the bursar.

7. Use commas to set off a noun of address. For example:

   What do you think, Trevor, about this solution?

8. Use commas to set off conjunctive adverbs such as *however* and *therefore*. For example:

   I understand, however, that we are responsible for damages.

9. Use a comma to set off tag questions. For example:

   Thaddeus is stupid, isn't he?

10. Use commas to separate certain items in dates. For example:

    July 1, 2007, is the termination date.

11. Use a comma to stand for non-repeated elements. For example:

    The red form goes to the bursar; the blue one, to the registrar.

12. Do not use commas to separate two independent clauses. This error, called a comma splice or comma fault, is one of the most serious in punctuation.

✗ POOR
John did not study, therefore, he did not pass.

✓ IMPROVED
John did not study; therefore, he did not pass.

✓ IMPROVED
John did not study. Therefore, he did not pass.

13. **Do not** use commas to separate triads of digits in metric usage. For numerals of five digits or more, use a space to separate triads on both sides of the decimal point. Four-digit numerals need no separation unless they are in tabular material with larger numerals.

INCORRECT
212,611; $1.20124; 2,600

CORRECT
212 611; $1.201 24; 2600

CORRECT IN TABULATIONS
2 600; 13 275

Dash

1. Use the dash to set off explanatory material that
   (a) has commas inside it
   (b) is separated from the noun or pronoun to which it refers, as in:

   John's book was widely read—a bestseller for months.

2. Use dashes sparingly, instead of parentheses, for emphasis.
3. Remember that putting a dash on one side of an item to set it off generally requires putting a complementary dash on the other side of the item, unless a period takes its place.

Ellipsis Marks

1. Use ellipsis marks (spaced periods) to show an omission within a quotation.
2. Type ellipsis marks to conform to the following:
   (a) Three spaced periods show an omission within a sentence.
   (b) Four periods show an omission that crosses over a sentence boundary: one period marks the end of the sentence, and the other three are the ellipsis.

Exclamation Point

Use the exclamation mark only occasionally, to *show strong emotion*.

Hyphen

1. Use a hyphen to separate the whole number from the fraction in a mixed number, as in *3-5/16*.
2. Use the hyphen in compound words, such as *self-actualization*.
3. Use the hyphen between compound modifiers, as in a *75-unit highrise* or *a computer-scored answer sheet*.
4. Use the hyphen in place names that include it, as in *Niagara-on-the-Lake, Ste-Anne-des-Monts*.
5. Do not use the hyphen
   (a) between words that merely follow each other

   COMPOUND MODIFIER
   20-dollar bills (= $20 x ?quantity)

   SEPARATE MODIFIERS
   20 dollar bills (= 20 x $1)

   COMPOUND MODIFIER
   A deeply-dredged canal

   ADVERB AND ADJECTIVE
   A deeply dredged canal

   (b) with an Arabic numeral and a metric abbreviation, even if you would hyphenate a parallel non-metric expression

   COMPOUND METRIC MODIFIER
   450 km drive

   COMPOUND NON-METRIC MODIFIER
   75-unit highrise

   (c) to break a word at the end of a line if
   (1) the entire word has fewer than seven letters
   (2) the division does not occur between syllables
   (3) the word is part of a proper name
   (4) the line is at the end of a page

(5) the hyphenated word comes immediately after or before another hyphen or dash
(6) the page has several other hyphenated words
(7) the manuscript is to be submitted to a publisher

Parentheses (Round Brackets)

1. Use parentheses for supplementary remarks or for references in the text, such as:

   (Joseph N. Ulman, Jr. and Jay R. Gould, *Technical Reporting*, 3rd ed. [New York: Holt, Rinehart and Winston, 1972], pp. 197–8).

2. If the sentence element before the parentheses requires punctuation, place the punctuation after the closing parenthesis.

   Ulman and Gould (pp. 197–8), while discussing parentheses, also cite this rule.

3. Use parentheses to enclose the area code for a telephone number: (504) 345-2063.
4. Use parentheses in pairs.

 POOR
1)

 IMPROVED
(1)

Period (Full Stop)

1. Use the period at the end of a declarative or imperative sentence or after a polite *would-you-please* request.
2. Use the period with certain abbreviations (U.S., f.o.b., P.D.Q.), but not with others (RCMP, YMCA, CNCP, metric designations). The use of the period is often optional, but should be consistent (if AM, then PM; if a.m., then p.m.).
3. Use the period for a decimal point.

Question Mark

1. Use the question mark for a direct question, but not for an indirect question:

   Judy asked, "Did I hurt someone's feelings yesterday?"
   Judy asked whether she had hurt someone's feelings yesterday.

2. Use the question mark in parentheses to express doubt about the material preceding the parentheses.

   John's new (?) car was a Model T.

3. Do not follow the question mark immediately with a comma, a period, or a semicolon.

Quotation Marks

1. Use double quotation marks to enclose a speaker's or writer's *exact* words.
2. If a quotation extends for more than a paragraph, use opening quotation marks before each quoted paragraph and use closing quotation marks only at the end of the quotation.
3. Use quotation marks to enclose the title of a work that is published as part of a book—for example, the titles of short stories, chapters, poems (see also "Underline (Underscore)," page 390).
4. If another mark of punctuation is immediately adjacent, position closing quotation marks as follows:
   (a) The comma or the period always goes inside (to the left of) the closing "quotation marks."
   (b) The colon or the semicolon always goes outside (to the right of) the closing "quotation marks":
   (c) The exclamation point or the question mark goes inside if part of the quotation, outside if part of the sentence surrounding the quotation.

   Mary said, "When will I see you again?"
   Did Mary say, "I will see you again tomorrow"?
   Did Mary say, "When will I see you again?"?

5. Use single quotation marks for quotes within quotes; inside the single quotes, revert to double quotation marks if necessary.

   John said, "Bill claimed, 'I have read the section entitled "Recognizable Patterns of Language."' "

6. Use double quotation marks to indicate words used in a special (sometimes satiric) sense. Be sparing with this device; do not use quotation marks as decorations.
7. Although quotation marks may be used in tabular columns to indicate repetition, do not use them for other abbreviations (for *inches* and *feet*, spell the words out or use the standard abbreviations *in.* and *ft.*).

8. When quoted material is separated from the text as a "block quotation," indented and single-spaced, use no quotation marks that are not in the original quote. The layout in itself indicates that the material is being quoted; quotation marks would, therefore, be redundant.

Semicolon

1. Use the semicolon to separate independent clauses that are closely related in logic.

   Semicolons are one thing; commas are another.

2. Use the semicolon to separate clauses or items in a series that have internal commas.

✗ POOR
Attending the meeting were Joseph Biggs, St-Jean, Quebec, president, David L. Carson, Kirkland Lake, Ontario, secretary, and Allen White, Duncan, British Columbia, treasurer.

✓ IMPROVED
Attending the meeting were Joseph Biggs, St-Jean, Quebec, president; David L. Carson, Kirkland Lake, Ontario, secretary; and Allen White, Duncan, British Columbia, treasurer.

See also "Comma," 4 (page 384), as in:

Suzanne, afraid of the final examination, studied frantically while drinking tea, coffee, and Pepsi Cola; and, in the end, she fell asleep during the test.

3. Use the semicolon between grammatically equal units only; do not use it, for example, between a dependent clause and an independent clause:

✗ POOR
While Sarah studied to be a CA; her boyfriend found a new girl.

✓ IMPROVED
While Sarah studied to be a CA, her boyfriend found a new girl.

Sentence Punctuation Error

Use only a period, a semicolon, a colon, a dash, or a comma plus coordinating conjunction to join two independent clauses:

	.	
	;	
	—	
This is the first	, and	This is the second
independent clause	, or	independent clause.
	, nor	
	, but	

See also Comma, 12 (p. 384).

Spacing

In typewritten work, be careful to use standard spacing rules:

1. Space twice after the period that ends a sentence.
2. Insert spaces between the dots in ellipses.
3. Space at least twice between the province and the postal code:

   The vendor is The Bible Shop, Post Office Box 267, Vancouver, BC V5Z 3J5.
   Mr. W.R. O'Donnell lectures in the Department of Linguistics and Phonetics, The University of Leeds, Leeds, West Yorkshire LS2 9CT, England.

4. Do not space before or after either the hyphen or the—(em dash).

Underline (Underscore)

Use underlining (underscoring) in typed or handwritten work for the same functions as italicizing in print:

1. Underline the titles of separately published works such as books and periodicals (see also "Quotation Marks", 3, page 388).
2. Underline expressions used as themselves.

   The word <u>word</u> should be underlined when used as a word.
   John's writing <u>32</u> instead of <u>23</u> in the last column caused the total to be $9 too high.

3. Underline foreign words and expressions that have not become fully anglicized.

The evaluator claimed that Ian's argument was <u>post hoc ergo propter hoc</u>.

4. Underline the names of vehicles, particularly ships.

David received valuable business experience while serving as disbursing officer aboard <u>Koutenay</u>.
A replica of the historic schooner <u>Bluenose</u> is on display in Halifax harbour.

Virgule (Slash)

1. Use the slash to indicate alternative possibilities.

Telephone Richard Sasseville at (524) 345-3682/2063.

2. Use the slash to indicate a fiscal or academic year that includes parts of two calendar years.

I shall complete my studies in 2006/07.
The 2006/07 financial year begins on July 1.

GUIDELINES FOR SPELLING, HANDLING NUMBERS, AND CAPITALIZING

Spelling

1. Spell words in accordance with a dictionary and accepted conventions. English spelling is more or less regular, but rules to describe it usually become submerged in exceptions. People who read a lot tend to spell well; it's partly a matter of pattern recognition. Some of the most troublesome words are listed in Table A.2.

    Notice that the table includes some alternatives. In writing English, Canadians can choose either the British or the American spelling conventions (for example, *colour/color*, *centre/center*, *defence/defense*, *programme/program*, *travelling/traveling*). In general, choose one convention or the other. For example, in this book we follow the *Nelson Canadian Dictionary of the English Language*.

2. Leave place names, organization names, and other proper nouns in their exact official forms—for example, *Canadian Radio-television and Telecommunications Commission*, not *Canadian Radio-Television and Tele-communications Committee*. Another example: notice the difference between *Saint John* (New Brunswick) and *St. John's* (Newfoundland). If you are using American-based spelling, remember that many Canadian proper nouns retain British forms:

TABLE A.2 Potential Spelling Problems

a lot	coolly	interpret	part-time	renowned
absence	cylinder	irresistible	pastime	repetition
accessible	decision	irritable	performance	rhythm
accidentally	definitely	laboratory	permissible	ridiculous
accommodate	definition	ledger	persistent	schedule
accommodation	define	leisure	personal	seize
accurate	describe	licence	personnel	sense
achievement	description	*or* license	picnic	separate
acquaintance	desirable	(noun)	picnicking	separation
acquire	despair	lose	possession	sergeant
among	development	loose	possible	sheriff
analogous	disappear	losing	practical	shining
analyze	disappoint	maintenance	precede	similar
apparent	disastrous	manoeuvre	predictable	studying
appropriate	discriminate	*or* maneuver	preference	succeed
arguing	drunkenness	marriage	preferred	succession
argument	efficiency	mere	prejudiced	suddenness
assistant	embarrassment	minuscule	prepare	superintendent
attorneys	environment	*or* miniscule	prevalent	supersede
balloon	equipped	misspelling	privilege	supposed
beginning	exaggerate	moral	probably	surprise
belief	exceed	morale	procedure	technique
believe	excellence	mortgage	proceed	than
beneficent	existence	necessary	profession	then
beneficial	existent	newsstand	professor	thorough
benefited	experience	ninety	prominent	through
carburetor	explanation	noticeable	pronunciation	tragedy
category	fascinate	occasionally	propeller	transferred
changeable	forty	occurred	psychology	tries
choose	government	occurrence	ptomaine	truly
chose	grammar	occurring	pursue	undoubtedly
coming	grievous	offered	quantity	unnecessary
committee	guarantee	offering	questionnaire	until
comparative	height	omitted	quiet	using
conscience	holiday	opinion	realize	vacuum
conscientious	imagine	opportunity	receipt	varies
conscious	immediately	original	receive	vicious
consensus	incidentally	paid	receiving	villain
consistent	indispensable	pamphlet	recommended	weird
controversial	insistent	panicky	referred	woman
controversy	interest	parallel	referring	writing
convenience	interface	paralyze	relevant	written

Labour Day, Department of Archaeology, Ministry of Defence. Proper nouns also retain their official punctuation—for example, write *Holt, Rinehart and Winston*, not *Holt, Rinehart, and Winston*.

Be particularly careful with French proper nouns; the conventions of punctuation and capitalization are somewhat different from English conventions: for example, *rue St-Jacques*, not *Rue St. Jacques*; *ministère des Transports*, not *Ministère des Transports*. But be careful; some French names use English conventions—for example, the *St. Boniface area of Winnipeg*.

3. Spell out the words for symbols unless you really need to save space:

NOT THIS
@
¢
%

BUT THIS
at
cent or cents
percent

Be especially careful with the # symbol. It has several meanings, and it is often redundant before a number. Why should you write *Apartment #45* when *Apartment 45* will suffice?

Numbers

Deciding whether to use words or numerals to express numbers in your business writing is not always easy. However, the following guidelines will help you make those decisions. Remember, above all else, to be consistent in your choice of words or numerals.

1. Spell out numbers from one to ten; use numerals for numbers above ten (11 and up). If numbers are in a series, treat them consistently. For example:

   John has two days' vacation left this year.
   His classes last for 50 minutes.
   This floor has 13 classrooms, 5 labs, and 21 offices.

2. Spell out a number that introduces a sentence. If the number is too large to spell out (1 376 984), reword the sentence so the number is not at the beginning. For example:

✗ POOR
50 students attended the meeting.

✓ IMPROVED
Fifty students attended the meeting.

✗ POOR
50 375 students attend the University of Toronto.

✓ IMPROVED
The University of Toronto has 50 375 students.

3. Spell out approximate numbers if they consist of only one or two words. However, such numbers can be written as numerals for emphasis. For example:

Approximately fifty thousand students attend the University of Toronto. More than 50 000 University of Toronto alumni have given to the alumni fund.

4. Spell out ordinal numbers (*first, second, third*, etc.).
5. Spell out fractions that are used alone (*one-half, two-thirds*, etc.); however, use numerals for mixed numbers ($11\ 1/2$).
6. Spell out large round numbers like million and billion for easier reading. For example:

Canada has a record deficit of $30 billion this year.

7. Use numerals for dates in most business contexts; notice that days of the month are written as cardinal numbers (*6, 12, 31*), even though they are spoken as ordinal numbers (*sixth, twelfth, thirty-first*). But spell out all numbers in formal invitations and announcements. For example:

LESS FORMAL
The meeting is scheduled for Monday, March 17.

FORMAL
On Monday, the seventeenth of March, you are invited ...

8. Differentiate two series of numbers by writing one series in words and the other in numerals. For example:

George had five 12-column ledgers and twelve 9-column ledgers.
I have five $20 bills and four $10 bills.

9. Use numerals, even for quantities less than 11, with all units of measure: money, percentages, time, dates, addresses, age, arithmetic calculations, metres, kilometres, inches, feet, stockmarket quotations, page numbers, volume numbers, degrees of temperature, and so on. For example:

He lives at 801 West Broadway.
He has grown 4 cm in the past month.

10. Express all numbers containing decimals as numerals, using a 0 (zero) where necessary to complete the notation. For example:

✗ POOR
He had a grade point average of three point five.

✓ IMPROVED
He had a grade point average of 3.5.

✗ POOR
He had a blood alcohol level of 1. when he had the accident.

✓ IMPROVED
He had a blood alcohol level of 1.0 when he had the accident.

Never use numerals only to the right of a decimal point.

✗ POOR
.75; .1%

✓ IMPROVED
0.75; 0.1%

11. Use a space, not a comma, to separate triads of numerals on either side of the decimal point (1 212 636.245 24). Four-digit numerals are usually left closed ($1236) unless they are tabulated with larger numerals.
See also "Hyphen," 1 and 5 (page 386).

Metric Units

Canada has adopted the International System of Units (SI) or metric units of measurement. When writing metrics, use the following conventions.[1]

1. Use only SI-approved units and their proper symbols (see Table A.3).

TABLE A.3 Some Common Metric Units and Their SI Abbreviations

Unit	Symbol	Unit	Symbol
centimetre	cm	megahertz	MHz
cubic centimetre	cm^3	metre	m
cubic metre	m^3	metre per second	m/s
degrees Celsius	°C	milligram	mg
gram	g	millilitre	mℓ or mL
hectare	ha	millimetre	mm
hertz	H	newton metre	N•m
kilogram	kg	square centimetre	cm^2
kilohertz	kH	square kilometre	km^2
kilometre	km	square metre	m^2
kilometre per hour	km/h	tonne	t
kilowatt	kW	volt	V
litre	ℓ or L	watt	W

2. Note the spelling of the names of the units:
   (a) *Metre*, *litre*, and their derivations take a final *-re* even if you are using the American convention that calls for *-er* at the ends of words such as *centre*.
   (b) Prefixes and base units take no hyphens (*kilogram*, not *kilo-gram*).
   (c) The names of all units are lower-cased except the modifier in *degree Celsius*.
3. Use either names or symbols; do not mix the two.

✗ POOR
newton m; N metre

✓ IMPROVED
N·m; newton metre

4. Always use symbols when you use numerals, and write out the names of the units when you write out numbers. In general, use numerals and symbols for exact quantities, even those less than 11:

✗ POOR
11 kilograms; six mL

✓ IMPROVED
11 kg; 6 mL

CORRECT
I drove a few kilometres down the road.

5. Mixed fractions are awkward with metric units; generally, convert them to decimal expressions:

✗ POOR
1 1/2 g

✓ IMPROVED
1.5 g

6. Metric symbols should be used accurately. Notice the following:
   (a) Most symbols are lower-cased. The exceptions are M (the symbol for the prefix mega) and the symbols for units named after people (for example, W, the symbol for watt).
   (b) The proper symbol for litre is a cursive (script) ell: ℓ. When this symbol is not available (as on many typewriters), you may substitute an uppercase ell: L. Do not use a lowercase ell—it is too easily confused with the numeral one.
   (c) Metric symbols never take periods.
   (d) Metric symbols do not change in the plural.
   (e) The proper symbols for squared and cubed are superscripts: km^2, cm^3.
   (f) Leave a space between numeral and symbol; the only exception is the degree symbol, which is closed so it won't get "lost":

✗ POOR
65km; 6 °C

✓ IMPROVED
65 km; 6°C

   (g) The symbols for compound units formed by dividing other units contain a virgule (slash). When these compound units are written out, however, use the word *per*.

✗ POOR
km per h; kilometre/hour

✓ IMPROVED
km/h; kilometre per hour

(h) The symbols for compound units formed by multiplying other units contain a dot signalling multiplication. The written-out forms of these compounds contain no special punctuation:

✗ POOR
kWh; kilowatt-hour

✓ IMPROVED
kW•h; kilowatt hour

7. In general, choose unit prefixes to avoid decimal fractions or to use their simplest forms.

✗ POOR
My height is 1.64 m.

✓ IMPROVED
My height is 164 cm.

✗ POOR
This brine shrimp is 0.00321 m long.

✓ IMPROVED
This brine shrimp is 3.21 mm long.

In general, you should choose a prefix that sets the numerical value between 0.1 and 1000. When a passage refers to similar quantities, however, use the same prefix for like items even if some values fall outside that range.

8. When you make a metric conversion for a business communication, consider whether precision or ease of comprehension will be more important to your reader. If, for example, a business deal involves four miles' worth of pipe line, you will want to discuss 6.43736 km. In ordering fencing, however, 6.44 km would suffice. For many purposes, 6 km would be close enough, and for others you could simply write "about 5 km."

Be sure the degree of precision you choose is sufficient for all comparable quantities in a passage. Like other numbers, comparable numerals with metric units should be carried to the same number of decimal places.

✗ POOR
The packages weighed 340.19 g, 425.242 g, and 567 g.

✓ IMPROVED
The packages weighed 340.2 g, 425.2 g, and 567.0 g.

Be aware, too, of whether your industry is making soft or hard conversions. Soft conversions retain the size used in the old Imperial system, with the measure given in metrics — often, of course, an unusual number. For example, a can that holds 14 fluid ounces may be relabelled 398 mL (notice that even this figure represents some rounding from the mathematical conversion of 396.89342 mL). Hard conversions change dimensions to even metric sizes. For example, the can itself would be changed to hold perhaps 400 mL.

See also "Numbers," 10 and 11 (page 395).

Capitalizing

1. In material to be capitalized, except in headings and other situations in which you use all-capitalization (ALL CAPS), capitalize the first letter of nouns, verbs, adjectives, and adverbs. Also capitalize the first letter of the first word in such material even if it is not a noun, verb, adjective, or adverb.
2. Capitalize the following:
   (a) Names of deities and of titles of scripture books.
   (b) Literary works such as books, articles, poems, and stories.
   (c) Important documents *(the Magna Carta, the Charter of Rights)*.
   (d) Days, holidays, months, and historical periods *(the Enlightenment)*.
   (e) The first word in a sentence.
   (f) Languages (see also 4b).
   (g) Organizations *(IBM, New Democratic Party, Roman Catholic Church)*, but preserve the capitalization that the organization uses officially (for example, *E.I. du Pont de Nemours & Company*, not *E.I. DuPont De Nemours and Company*).
   (h) Places and regions *(the Maritimes, the West)* (see also 4c).
   (i) Names of streets and other thoroughfares.
   (j) Title when followed by a name *(Miss Alice Young, Professor Yaney)*.
   (k) *Father, Mother, Brother, Sister*, etc. when used like a given name, as in the salutation or greeting of a friendly letter ("Dear Mom").
   (l) *Prime Minister* whenever used with reference to the head of the Government of Canada. Note, too, the capitalization of *Government* in this context.
   (m) Certain nouns if followed by numbers, as in *Apartment B727*.
   (n) Nouns intended to stand for an entity that would be capitalized. (*College*, for example, would be capitalized if intended to stand for Atkinson

College of York University, but not if used in the sense of *Joe went to college.*)

3. Be consistent in capitalization of comparable words.

INCORRECT
The class contained both anglophones and Francophones.

CORRECT
The class contained both anglophones and francophones.

CORRECT
The class contained both Anglophones and Francophones.

CORRECT
The officers present were Bonnie Campbell (President) and Pat Underwood (Treasurer).

CORRECT
The officers present were Bonnie Campbell (president) and Pat Underwood (treasurer).

4. Use lowercase (small) letters for
   (a) seasons of the year (*winter*)
   (b) academic subjects not otherwise capitalized:

   Brenda, Dee, and Terry respectively studied management, Roman history, and English.

   (c) simple directions, as on a compass (see also 2h)

   After spending the summer in the north, John moved south in October.

5. Check for correct usage on all French proper nouns.

Notes

Richard David Ramsey, of Haliburton Services, Duncan, OK, prepared this Appendix, which has been adapted for this edition.

---

[1] For more information, see the *Canadian Metric Practice Guide*, CSA Standard CAN 3-Z234. 1-79 (Rexdale, ON: Canadian Standards Association). Another aid is *The Metric Guide*, 2nd ed. (Toronto: Council of Ministers of Education, Canada, 1976).

Index

A/an confusion, 366
Abstract language, 27–29
Activities checklist, 202
Addresses, format of, 67
Adjectives, 362, 366–67
Adverbs, 362, 366–67
Agendas
    of meetings, 185, 186
    of teleconferences, 191
AIDA sequence, 149–51
    in application letters, 229–32, 234–37
    in complaints, 161–63, 170
    in oral communication, 167–68
    in special requests, 158–60
Analytical reports, 281, 305, 315–19
APA (American Psychological Association) style, 293, 294–96, 297–98
Apologies, 37, 111, 113, 131, 133, 135
Apostrophes, 381–82
Appearance, personal, 14, 147
Appendices, 337
Application forms, 239–40
Application letters, 229–33, 234–37
Asynchronous technologies, 193
Audience, 7, 8, 10–12, 15, 167–68, 177, 180

Bad news messages, 89, 127–40
Bar graphs, 352, 353
Belonging needs, 153, 154
Bibliographies, 293, 295, 296
Body language, 8, 14, 181
Books, 282–83
    citation of, 295, 296
Brackets, 382
Bulletin boards, web-based, 193, 237, 239
Buzzwords, 355

Capitalization, 399–400
CD-ROMs, 283
Citations, of research sources, 292–93, 295, 296
Clarity, 22–31
Clauses, 363–64, 367
Clichés, 355
Colons, 382–83

Commas, 383–85
Communication
    barriers, 9, 15
    business communication style, 22–46
    medium selection, 10–12, 15
    non-verbal, 181
    in social *vs.* work life, 4
    supervisory, 163–67
    technological, 191–93
    See also Oral communication; Written communication
Complaints, 161–63, 164
    oral communication and, 170
    responses to, 110–18, 130–38
Conciseness, 23–25, 30
Concrete language, 27–29
Conjunctions, 362–63
Courtesy titles, 42–43, 57, 59
Co-workers, qualities of, 205
Credibility, 147
    of opinions, 270
    of reports, 274
    of websites, 290

Dangling expressions, 368
Dashes, 385
Databases, 284, 285, 287
    articles, citation of, 295, 296
Dates, format of, 394
Decimal fractions, 395, 397, 398
Decision-making, 148–49
Decoding, of messages, 3
Deduction, 271, 272
Dependent clauses, 364
Dependent variables, 352
Direct order
    in good news messages, 108, 110, 118
    in reports, 306, 317, 331
Documentation of sources, 293–98

Editing, of formal reports, 355–56
Ellipsis marks, 385
E-mail, 10, 11, 69–71, 255
Emotions
    complaints and, 162, 170

role in decision-making, 148–49
Employers, job interviews and, 250–53
Employment. *See* Jobs
Encyclopedias, 283
English language
    grammar, 29–30, 221, 356, 360–65
    punctuation, 29–30, 60, 381–91, 395
    spelling, 29, 221, 356, 391–93
    usage, 365–81
ESL tips, 27, 60, 115, 137, 220, 229
Esteem needs, 153, 154, 159
Ethnic origin, 41–42
Exclamation points, 386
Explanations
    in bad news messages, 128, 132–34
    in complaints, 170
    in good news messages, 109, 111
Expletives, 363
Eye contact, 14, 181

Facial expression, 181
Facts, 268, 273, 291–92
    in decision-making, 148
Fax messages, 237, 255
Feedback, 8
    from interviews, 255–57
    negative, 182, 255–57
    in project groups, 189
Feelings. *See* Emotions
Follow-up
    to application letters, 232
    to job interviews, 253–56
    questions, 280
Fonts, 56, 165
    See also Typefaces
Formal reports, 331–56
    See also Reports
Forums, web-based, 193
Full block format, of letters, 61–62, 63

Generalizations, 271–72
Gerundive phrases, 363
Gerunds, dangling, 369
Gestures, 14
Good news messages, 108–20
Goodwill
    closings, 92, 109–10, 115, 119, 170
    complaints and, 163
    telephone requests and, 99
Government documents, in research, 284
Grammar, 29–30, 221, 360–65

Grammar checkers, 356
Graphics, in reports, 350–55
Graphs, 180
Groups
    closed user, 285
    project, 187, 189–91
    small, 178–81, 184–91

Hand-outs, in oral presentations, 180
Harbrace Handbook for Canadians (Hodges and Stubbs), 360
Headings, in reports, 315–17, 320, 354
Hyphens, 386–87

Incident reports, 307–8, 309
Incomplete constructions, 369
Independent clauses, 364
Independent variables, 352
Indexes, to periodicals, 284
Indirect order
    in bad news messages, 128, 139
    in reports, 306, 317, 331
Induction, 271
Industries, information about, 278
Inferences, 271
Infinities, split, 374
Infinitive phrases, 363
Informal reports, 305–24
    chronological, 307
    formal *vs.*, 331
    functional, 307
    See also Reports
Informational reports, 281, 305–15
Intensifiers, 362
Interjections, 363
Internet, the
    Canadian websites, finding, 288
    directories, 287
    e-mail, 10, 11, 69–71, 255
    government websites, 284
    inquiries via, 156
    intranets, 193
    job applications and, 237, 239
    job search and, 227–29
    library collections, 282
    meta search tools, 287–88
    orders via, 156
    periodicals, 283
    search strategies, 286–91
    secondary research and, 282, 283, 284, 285–91

web conferencing, 191, 192–93
website addresses, 286
website citations, 295, 296
website evaluations, 290
Interviews
    job, 232, 239, 248–58
    reference page and, 296
    telephone, 277–81
Intranets, 193

Jargon, 26
Job market, researching, 227–29
Jobs
    advertisements/postings, 234–39, 241
    applications, 237, 239, 241–42
    employability skills, xiii
    interviews, 232, 239, 248–58
    searching for, 201, 227–42
Journals
    citation of articles, 295, 296
    professional, job search and, 227, 228

Key word searches, 282

Letters, 55–67
    of application, 229–33, 234–37
    elements of, 56–60
    formats of, 61–63
    headings, 56
    reaching audience with, 11
    salutations in, 57–58
    spacing of, 55–56
    stationery for, 55, 56, 63, 66
    transmittal, 332–33, 334
Libraries, research in, 282–84
Line graphs, 352, 353
Listening, 9, 192
Lists, 82, 350
Listservs, 286
Logic, in decision-making, 148
Logical conclusions, 270–74

Magic words, 151
Mail orders, 156
Maslow, Abraham, 153
McLuhan, Marshal, 8
Meetings, 184–91
    agendas of, 185, 186
    chairpersons, 185–86, 187
    conducting, 185–86
    format and length of, 185

minutes of, 185, 187, 188, 281
participation in, 187
productivity of, 185
Robert's Rules of Order, 185
teleconferencing, 191–92
typical problems, 190–91, 193
videoconferencing, 192
web conferencing, 192–93
Memoranda, 68–69, 332–33
Messages, 3
    audience of, 7, 15
    bad news, 89, 127–30
    delivery medium, choice of, 10–12, 15
    e-mail, 10, 11, 69–71, 255
    good news, 108–20
    non-verbal, 12, 14
    oral *vs.* written, 95
    persuasive, 147–70, 230–31, 235
    planning of, 4, 13, 15
    purpose of, 4, 15
    revision of, 10
    routine, 80–95
    sales, 151–57
    unsuccessful, 15
    voice mail, 10, 11
Metric units, 395–99
MLA (Modern Languages Association) style, 293, 294, 296
Modified block format, of letters, 62
Modifiers, 362
    misplaced, 369–70

Needs, human, 153–54
Negativity
    in bad news messages, 130, 134
    in body language, 14
    of feedback, 182, 255–57
    of tone, 35–40
Nervousness
    at job interviews, 248, 249, 252
    during presentations, 177, 182, 183, 184
    during tests, 257
Networking, job search and, 228
News groups, 286
Non sequiturs, 272–73
Non-verbal communication, 12, 14, 181–82
Note-taking, 281
Noun clauses, 367
Noun phrases, 363
Nouns, 361
    possessive case of, 381–82

proper, 391, 393
See also Grammar
Numbers, 393–95

Objective language, in reports, 268, 274
Opinions, 269–70, 273, 292
    expert, 292
Oral communication, 8–12
    AIDA sequence in, 167–68
    bad news messages and, 139–40
    clarity in, 30–31
    complaints and, 170
    good news messages and, 118–20
    importance of, 176
    job applications and, 241–42
    persuasion in, 167–70
    presentations, 176–84
    routine requests and, 95–99
    tone in, 44–47
    visual impact in, 147
    See also Meetings
Orders
    handling telephone requests, 97–99
    placement of, 85–89, 96–97
Overhead transparencies, 183

Paper
    for formal reports, 354
    for letters, 55, 56, 63, 66
    for résumés, 220
Parallelism, 370
Paraphrasing, 292
Parentheses, 387
Parenthetical citations, 294
Participial phrases, 363
Participles, dangling, 368
Passive, the, 370–80
Periodicals, in research, 283–84
Periodic reports, 312, 314
Periods (punctuation), 387
Personal qualities, 204–5
    job advertisements and, 234
Persuasion
    in oral communication, 167–70
    persuasive messages, 147–70
    persuasive reports, 305, 319–24
    See also Promotions; Sales
Photocopying, 183
Phrases, 363
Pie charts, 354–55
Plurals, foreign-language, 372

Positiveness
    of closings, 113, 134
    in presentations, 182
    of tone, 35–40
Posture, 14, 181
Prepositional phrases, 363
Prepositions, 362, 371
Presentation software, 178–81, 184
Presentations, oral, 176–84
Primary research, 274, 275–81
Process, defined, 3
Progress reports, 308, 310–11
Projects, 187, 189–91
Promotions, tone in, 38–39
Pronouns, 361–62, 371–73
Pronunciation, 31
Proofreading, 366
    reports, 356
    résumés, 221
Proper nouns, 391, 393
Proposals, written, 319–24
Provinces and territories, symbols and French names of, 67
Provocative statements, 151
Public speaking, 176, 183, 184
    See also Presentations, oral
Punctuation, 29–30, 381–91
    mixed, 62
    of numerals, 393
    open, 62

Question marks, 387–88
Questionnaires, 275–77, 337
Questions
    closed, 276, 279–80
    factual, 279
    follow-up, 280
    in interviews, 252–53, 279–80
    main, 267
    open-ended, 276, 277, 280
    opinion, 279
    in oral communication, 96
    research, 267, 275
    in sales messages, 151
Quotation marks, 388–89
Quoting, from sources, 293

Racism, 41–42
Random samples, 276
Reading, listening *vs.*, 9
Receivers, 3

feedback from, 8
of letters, 57
See also Audience
Recommendations, in formal reports, 331, 336
Records, permanency of, 9–10
Redundancy, 23–24
Reference books and CD-ROMs, 283
References, 293, 295, 296, 336
References, on résumés, 219
Repetition, 23, 170, 178
Replies, routine, 89–95
Reports
    analytical, 281, 305, 315–19
    appearance of, 331, 333, 335, 354
    conclusions of, 336
    direct/indirect order in, 306, 317, 331
    formal, 331–56
    incident, 307–8, 309
    informal, 305–24, 331
    informational, 281, 305–15
    periodic, 312, 314
    persuasive, 305, 319–24
    on printed forms, 307, 314–15
    progress, 308, 310–11
    recommendations in, 331, 336
    sales-call, 281
    trip, 312, 313
Requests
    routine, 80–85, 95–99
    special, 158–60
Research
    interviews, 277–81
    for job applications, 251
    primary, 274, 275–81
    purpose of, 267
    secondary, 274, 281–91
Résumés, 201–22
    appearance and accuracy of, 220–21
    application forms *vs.*, 239
    chronological, 208–13
    defined, 208
    educational experience, 208–9, 210, 212, 215
    electronic, 221–22
    employment objectives, 215, 219
    faxing, 237, 255
    functional, 209, 214–20
    hobbies, 212, 250
    identification information, 210, 214
    job interviews and, 250
    posting on web-based bulletin boards, 237, 239
    preparation of, 208–21
    pre-resume analysis, 201–22
    proofreading, 221
    references, 219
    samples, 211–12, 213, 216–18
    skills format of, 209, 214–20
    social insurance numbers and, 210
    telephone calls and, 241
    updated résumé, sending, 255
Robert's Rules of Order, 185

Safety needs, 153, 154, 165
Sales
    messages, 151–57
    personal, 156
    reports on calls, 281
Sales-oriented language, 91
Salutations, in letters, 57–58
Secondary research, 274, 281–91
Security needs, 165
Senders, 3
Sentences
    fragments of, 373–74
    structure of, 29, 364
Sexual bias, avoiding, 42–44, 372
Signature blocks, of letters, 59
Skills, 202–3
    employability skills, xiii
    on résumés, 209, 214–20
Slashes, 391
Slides, in oral presentations, 178–81
Small groups, 178–81, 184–91
Sources of information, 274–75, 278, 291–98
Spacing
    of letters, 55–56
    in writing, 390
Speaking, 8–9, 11–12
Special requests, 158–60
Spell checkers, 356
Spelling, 29, 221, 391–93
Split infinitives, 374
Statistics, 337, 352
Statistics Canada, 284
Subheadings, 316–17
Subjective language, in reports, 269, 274
Subject lines
    of application letters, 236
    of memoranda, 69, 71
    of reports, 308, 312

in supervisory communication, 164–65
Subjects, agreement with verbs, 374–76
Summaries
   of formal reports, 335–36
   of source information, 292
Supervisory communication, 163–67
Supporting data, 337
Symbols, 393
   metric, 396, 397–98
Synchronous technologies, 193

Tables
   in reports, 337, 351
   in routine requests, 82
Tables of contents, of reports, 335
Technological communication, 191–93
   See also Internet, the
Teleconferences, 191–92
Telephone
   complaints, 170
   follow-up calls, 232
   inquiries, 95–99, 156
   interviews, 277–81
   job applications by, 237, 239, 241–42
   messages, 11, 95–99
   note-taking from, 281
   orders, 156
   in research, 277–81
   voice mail, 10, 11
Tests, standardized, 257–58
Thank-you, in routine requests, 83
Title pages, of reports, 335
Tone
   in communication, 31–46
   non-discriminatory, 41–44
   in oral communication, 44–47
   of voice, 8, 14, 46
   you-centred, 32–35
Trade shows, job search and, 228
Transition words, in oral presentations, 181
Transmittal memos and letters, 332–33, 334
Transparencies, 183
Trip reports, 312, 313
Typefaces, 56
   of formal reports, 354
   for résumés, 220
   in sales messages, 151

ULR (Uniform Resource Locator), 286
Underlining (underscoring), in writing, 390–91

Variables, 352
Verb phrases, 363
Verbs, 362
   agreement with subjects, 374–76
   tenses, 376–77
   See also Grammar
Videoconferences, 192
Virgules, 391
Visual images
   in oral presentations, 178–81
   in sales messages, 151
   in supervisory communication, 165
Voice
   in presentations, 181–82
   tone of, 8, 14, 46
Voice mail, 10, 11

Web conferencing, 191, 192–93
Websites. See Internet, the
What Color is Your Parachute? (Bolies), 208
Wordiness, 24–25, 377–78
Words, 361–63
   abstract, 27
   choice of, for clarity, 30–31
   concrete, 27–29
   magic, 151
   misuse of, 377–80
   sexist, 42–44, 372
   simple vocabulary, 25–27
   transition, 181
   See also English language
Work
   conditions of, 206–7
   experience, on résumés, 208–9, 210, 215
   rewards from, 207–8
   values, 207–8
Works Cited, 293, 296
Written communication, 8–10, 11–12, 95
   bad news messages, 127–38
   business style, 22–30
   ESL tip, 27
   good news messages, 108–18
   proposals, 319–24
   routine messages, 80–85
   visual impact in, 147

You-centred communication, 32–35, 150, 165
   in presentations, 182